Living a Vineyard Dream

Clifford Ehrlich

Printed in the United States of America

ISBN-13: 978-0-9716705-8-7
ISBN-10: 0-9716705-8-7

Library of Congress Control Number: 2016907230
Living a Vineyard Dream

To contact Clifford Ehrlich or the Ehrlich Vineyard,
send an email to ehrlichvineyard@aol.com

Published by Legacy Family History, Inc.
Highland, Utah

Dedicated to

*Patricia, my co-adventurer and loving companion of a lifetime,
and Susan, Brian, and Scott*

"Nothing happens without first a dream."
–Thomas Edison

"Dream no small dreams.
They have no magic to move men's souls."
–Thomas Bentham

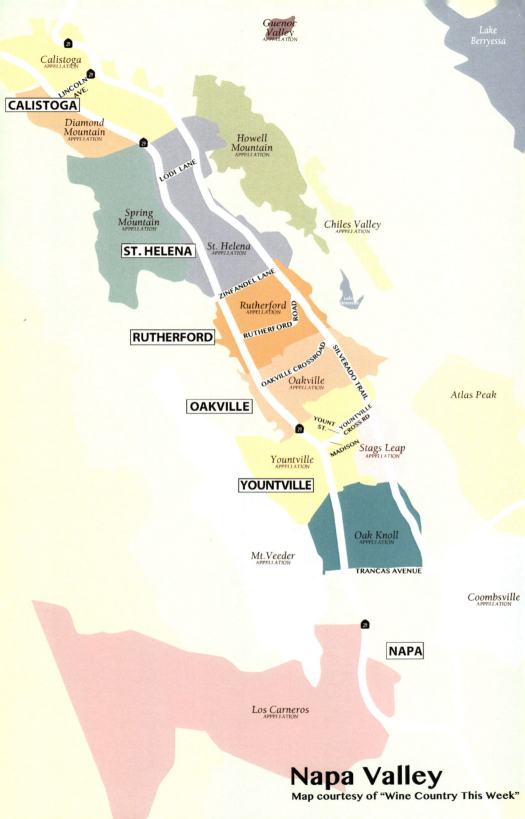

Lake
Berryessa

Guenoc
Valley
APPELLATION

29 *Calistoga*
APPELLATION

29

CALISTOGA

LINCOLN AVE.

Diamond
Mountain
APPELLATION

29

Howell
Mountain
APPELLATION

LODI LANE

Spring
Mountain
APPELLATION

Chiles Valley
APPELLATION

St. Helena
APPELLATION

ST. HELENA

ZINFANDEL LANE

Lake
Hennessey

Rutherford
APPELLATION

RUTHERFORD ROAD

RUTHERFORD

OAKVILLE CROSSROAD

SILVERADO TRAIL

Oakville
APPELLATION

Atlas Peak

OAKVILLE

YOUNT ST.

YOUNTVILLE CROSS RD

29

MADISON

Stags Leap
APPELLATION

Yountville
APPELLATION

YOUNTVILLE

Oak Knoll
APPELLATION

Mt. Veeder
APPELLATION

TRANCAS AVENUE

Coombsville
APPELLATION

29

NAPA

Los Carneros
APPELLATION

Napa Valley

Map courtesy of "Wine Country This Week"

Contents

Acknowledgements

I was born in New York City (Queens, not Manhattan) and grew up in New Jersey, but acquired a fascination for the West at an early age. My first trip there was in 1959, during the summer before my senior year in college. My exposure to the Rockies, California, and the desert left me wanting to see more.

My first job after graduation took me to the Boston area where I met my wife Patricia. She and I jumped at the chance to accept a transfer to California when it was offered. It was while we were exploring the state a year later that we discovered the Napa Valley. My internal gyroscope kept sending me there even after we moved back east again, and I instinctively knew one day it would become part of my life.

When we eventually made the decision to buy real estate there, the person who was most helpful was my realtor, Steve Ericson. Meeting Steve was serendipity—pure chance and good fortune. I tip my hat to him for his wise advice and counsel. His insights about the valley and the people who live there were invaluable. The gusto, spirit, and grit with which he greets every day make the time spent with him interesting and worthwhile. He continues to be a valued friend.

Laurie Wood, Tom and Aaron Pulliam, Guy Larsen, Frank D'Ambrosio, and Becke Oberschulte were invaluable in helping us settle into our new surroundings. Each generously contributed their time, special skill, and know-how to our well-being. They made the transition an easy one.

Of course, this story could not have occurred if Patricia was opposed to owning a vineyard. When she realized how important it was to me, she became devoted to the idea. Without her support, I would not have tackled this adventure, since no marriage is likely to survive the strain of an act as foreign as this was to the preceding thirty-five years of our life together.

The words of encouragement provided at appropriate times by our children, Susan, Brian, and Scott, were reassuring. They have found the vineyard as enjoyable as I do. I'm glad my mother, my sister Elaine, and Elaine's husband George had the chance to visit, and regret that my father and I were never able to walk together among the vines.

I'm grateful to the people who helped me write this book, particularly Laurie Wood, Jamie Davies, Jim Hickey, Ed Weber, and Jeff Popick, who all agreed to be interviewed for it. Each shared stories, experiences, and insights that helped me understand the Napa Valley and how it developed into the place it is today. In addition to being interviewed, Jim took time to read and edit my version of the creation of the Agricultural Preserve that appears in Chapter 6. In all cases, I have related what they told me as accurately as I know how.

There are a number of excellent books and articles written about the history of the valley. I read many of them and found myself relying on two almost exclusively while writing Chapters 9 and 10. They are *Napa Wine— A History* by Charles I. Sullivan (San Francisco: The Wine Appreciation Guild, 1994) and *California's Napa Valley—One Hundred Sixty Years of Wine Making* by William Heintz (San Francisco: Scottwall Associates, 1999). Messrs. Sullivan and Heintz did a wonderful job of researching and writing, and I am indebted to them for the facts they uncovered and their prose. The information about the geology of the valley in Chapter 4 was gleaned largely from an article written by John Livingston titled "The Geology of Fine Wine" that appeared in the Summer 1999 issue of the *Napa Valley Wine Library Report*.

My vineyard search occurred as I was wrapping up a business career that spanned almost forty years, most of it spent with Marriott International. I worked for Bill Marriott for twenty years, and it would be difficult to imagine a more dedicated, principled, or competent business executive than he. The success of the company he leads is a direct reflection of his talent, energy, and wisdom. It is highly unlikely that he would ever want to own a vineyard but, being the person he is, I know he shares my enthusiasm for the one I bought.

Despite the generous help I received from others while writing this book, the final product is mine and I take responsibility for any errors or misstatements.

It has been as much fun looking back at the events recounted here as it was to live through them.

Chapter 1:
The Vineyard Doctor Pays a Call

A blue Ford pickup slowed at the top of the gravel driveway and rolled to a stop beneath the trees. The hinge groaned as the driver pushed open the door, got out, and walked over to the well that was located in the shade of overhanging branches. The green canopy above him barely rustled as a squirrel tightroped his way along a low limb and disappeared into the thick foliage.

He knelt on the concrete pad and scanned the network of tanks, pipes, valves, and filters. Spitting on his thumb, he tried to rub years of accumulated rust and dirt from the tags on them. He squinted, hoping to find numbers or words that might be useful, but corrosion and old age had left nothing he could decipher. His hand swatted unconsciously at a fly as he stood up, returned to the truck, and drove the remaining two hundred feet to the parking area in front of the house.

"So you're the new guy with the water problem," he said to me lightheartedly, a smile spreading across his lined face. "I'm Laurie Wood. We've talked on the phone."

He wore jeans and a long-sleeved shirt with red and white vertical stripes. His straw hat was frayed at the brim and tilted back on his head. It was a relief to finally meet the man whose firm, calloused hand I was shaking. I had called him a week earlier.

He turned slightly and surveyed his surroundings. The two-story farmhouse had a wrap-around porch with an orchard in front and vineyards behind it. In a garden area just past the driveway, rose bushes displayed their blooms of white, yellow, red, and pink. A stretch of lawn led to the house where the colorful green and yellow leaves of variegated viburnum and assorted other plants bordered the porch. The outline of hills to the east, north, and west left no doubt that this was the middle of the valley.

"Haven't been here for awhile," he added. "Nice place. I knew the former owner. We had some dealings. Place hasn't changed much."

"I looked for two and a half years to find it," I said. "Patricia and I visited the valley in 1967 for the first time. We've been coming back regularly ever since."

It was a cautious comment. Laurie might not like newcomers. I didn't want to be mistaken for one of the Johnny-come-latelys buying land in the Napa Valley who was too young to spell "grapevine" in the '60s.

"It was different back then," Laurie observed.

"Sure was. I remember eating at the Grape Vine Inn before it got resurrected as Brix."

He laughed. "That was the best of a sorry lot. It didn't have any real competition. There's been a lot of growth around here since then in everything—restaurants, people, wineries. I've farmed vineyards since the late '40s and there's more business now than ever. It's humming. I couldn't be busier. How can I help?"

I explained that an analysis of my well water showed a high boron reading. The level was safe for people but not grapevines. Laurie speculated that a new well might be the answer, but first he wanted to look at the vines. As we headed toward them, he pointed to one of the three-dozen walnut trees that occupied the front few acres of the property, the remnants of an orchard that once occupied the land. Spaced fifty feet apart in parallel and perpendicular rows, they were a grafted variety that had grown to a height of thirty feet. The highly textured, dark-brown bark on the lower trunk contrasted with the streaked, silvery branches that seemed to burst upward from the graft to create the impression of energy and grace.

"Those are Hartleys," Laurie observed. "From the looks of it, they've been here for seventy or eighty years. Probably planted during Prohibition. We had lots of walnut and plum orchards back then, but most were taken out to make room for grapes. This is prime vineyard land. Gonna take yours out?"

"No. I'd rather keep a link to the past."

Laurie nodded his agreement and continued across the clumps of dry, grayish-brown loam that a tractor had left in its wake weeks earlier. As we walked, he mentioned some of the ailments that afflict walnut trees and emphasized their need for attention. Now standing with his hands on his hips, he surveyed the Sauvignon Blanc vineyard that was the cause of my concern. The coolness of early morning had vanished, and with the July sun almost overhead, the temperature was approaching the low eighties. The brim of Laurie's hat cast a shadow over his face, and he wiped a few beads of sweat off his chin with the back of his hand.

"When were they planted?" he asked, looking out over three acres of scrawny, gangly vines struggling up the stakes that supported them. They were pencil thin and with only a few leaves on each looked frail and undernourished.

"Don't know for sure. A year ago. Maybe two."

"They should look better this time of year," he remarked, and he walked to the nearest row to get a closer look. He explained that the dark-brown color along the perimeter of some leaves was an indication of high boron and pointed to others showing the same symptoms. He continued along, stopping periodically to inspect both the leaves and the vines they sprouted from. He was at home in a vineyard and had the manner of a doctor walking into a hospital ward, looking at the patients and reading their charts—observing, touching, grunting, and nodding. At the end of the row, he just shrugged.

"Those are unhealthy, all right. That's the bad news. The good news is that you have to work hard to kill grapevines. They're hardy. It's not too late to save them."

He shifted his attention to the thick trunks and sprawling canopy of four acres of Chenin Blanc that lay behind the Sauvignon Blanc, separated by twelve feet of open space. From their size, he estimated they were twenty or twenty-five years old. He explained that the absence of irrigation lines meant the soil was rich and recommended that we stop irrigating the Sauvignon Blanc. "Letting it grow without water is better than poisoning it with bad water."

As he moved between the rows of Chenin Blanc, he looked at the leaves and the cordons that spread from the main vine. He stopped from time to time and pointed to scars and marks and thoughtfully announced, "This is a mess. It's got just about every ailment a vineyard can get."

He cupped in his hands a cluster hanging from a gnarly vine. "This is bunch rot," he declared, looking at grapes that were mushy and others that had dried up, turned black, and looked like raisins.

He described other evils that were evident on the vines, pointing out examples of eutypa, Pierce's disease, and phylloxera, and taking the time to explain each of them. While the information was instructive, it was laced with words and terms that were unfamiliar to me, and my memory was fogging over. I mentioned that my vineyard manager had recommended that the vines be pulled out after harvest.

"That's the best advice he could give you," Laurie stated firmly. He knew the manager and added, "He grew up in the valley and worked for me when he graduated from UC Davis. He knows the vineyard business real well. Listen to him."

Now at the rear of the property, we followed the perimeter

of the vineyard back to the house and chatted about what we had seen. The muffled sound of farm equipment moving among the rows of an adjacent vineyard provided the backdrop for our conversation. A light breeze caused the vines to sway and provided a refreshing puff of coolness. I raised the idea of drilling a new well.

"You should have one," Laurie said. "Even though the water table here is pretty high, you'll still benefit from a good well. A vineyard does best with regular irrigation during the first few years."

Reaching the house, he cut across the orchard to his truck and explained, "What I do is called water witching or water dowsing. Most people don't believe in it. Geologists say it's bunkum, but that doesn't bother me. I've got a feel for finding water in this valley, and I've found it for a lot of people."

He leaned into the cab and removed two L-shaped copper rods. One side looked to be about three inches long, and the other was three to four times longer. Grasping one in each hand by the short side with the long ends pointing forward, he walked toward the well and watched as the ends began to move toward one another. "That means there's water. Whoever drilled here knew what they were doing."

He made the rod in his right hand whip up and down and counted the number of times it oscillated before coming to rest. Removing a small notebook and pencil from his breast pocket, he jotted something down. From the pocket of his jeans, he took a small, smoked glass bottle of liquid with a string tied around its neck, spun it around, counted the number of twirls it took before coming to rest, and made another note.

"It looks like the flow from your well is 55 to 65 gallons per minute," he remarked. "That's about all I can do today. I'll come back early next week to walk the property and pick some drilling sites. Think about two wells so you'll have enough for irrigation and frost protection. You haven't mentioned frost protection, but consider it. Overhead sprinklers are the best, and work on the same principle as the sprinklers you see in orange orchards in Florida. You need 50 gallons per minute per acre to operate them. That's 350 gallons per minute for your seven acres. We might be able to get that much water from two wells. In the meantime, don't worry. You've got a good spot."

As he got into his truck, he suggested the Chenin Blanc clusters be thinned as soon as possible. "Those vines are so sick they'll struggle to ripen the fruit that's on them. You stand the best chance of getting some decent production if you let them concentrate their energy on fewer clusters."

He waved as he drove off, and I returned to the cool of air conditioning, realizing Laurie had just put me through the introductory class of Vineyards 101. He filled my head with new information and vocabulary that were going to be part of my new life. I jotted down everything I could recall, but already had lost some of the details. Twisting off the top of a bottle of water, I took a long drink and remembered his estimate of the well's output. I shuffled through a stack of documents to find the results of the flow test that had been done professionally a few weeks earlier. I had the only copy. It showed 55 to 70 gallons per minute. That made Laurie either a world-class guesser or someone with an uncommon skill.

Chapter 2:
The Spark Gets Lit

The circumstances that brought me face-to-face with Laurie Wood had begun to unfold almost three years earlier in October 1995 when Patricia and I were visiting the Napa Valley. Harvest time is when the valley is most picturesque, and it was already in full swing when we arrived. The picking starts at the south end in Carneros with grapes used to make sparkling wine and moves progressively northward. One by one, the grape varieties ripen—Sauvignon Blanc, Chardonnay, Pinot Noir, Merlot, Zinfandel, and Cabernet Sauvignon—in Napa, Stags Leap, and Yountville; then on to Oakville, Rutherford, and St. Helena; and finally up to Calistoga. It is the annual agricultural explosion that creates a tourist overload for valley residents.

We were at dinner with Tom Paine and Teresa Norton, friends I had met several years before when our business careers had intersected. They had retired to the valley a few years earlier and created Vineyard 29, a boutique winery that would soon release its first vintage of Cabernet Sauvignon. When I expressed interest in becoming a Napa Valley landowner, they shared their experience of looking for property.

"We used Up-County Realtors in St. Helena," Teresa volunteered. "They understand the market and steered us here," gesturing with her hand to our surroundings. "Prices cratered right after we bought, but they seem to have turned the corner."

"The valley really felt it when the California economy hit the skids a few years ago," Tom explained. "We had the shock of watching our newly acquired real estate shrink in value at the same time we were laying out money to build the house and plant the vineyard. It's been a financial drain. The little rebound we've seen in land prices is a welcome relief."

Their words were sobering, yet Patricia and I saw no danger signs in the view from the patio of their contemporary home perched on the valley's western slope just north of St. Helena. In all directions lay a beckoning landscape of well-tended vines that rolled across the valley floor and decorated some hillsides. It was a classical pastoral setting that drew you into it.

The realtor we contacted the next day got the ball rolling. Among her listings were half a dozen properties she thought might suit us. Three looked interesting. The first two were houses with small vineyards. Both, however, were rundown and lacked any trace of curb appeal. The third was an early twentieth-century house on just over twelve acres. Located on Highway 29 about two miles north of St. Helena, it offered some potential.

"It was listed in March," the realtor explained. "It's been in the family for a long time. I don't know whether they've had many lookers or how flexible they are on the price. I hear they recently turned down an offer pretty close to the asking price."

To my uneducated eye, it seemed like a lot of money for what was there. The acreage was not planted and there was nothing to indicate whether it would be suitable for a vineyard. While the puffy language of real estate portrayed the house as quaint, it would have been more accurate to say it was frail, in need of repairs, and sagging under the weight of its years. A calculation of what it would cost to buy and renovate it and then plant a vineyard quickly reached a figure that exceeded our interest. But the main reason we weren't interested had nothing to do with cost. We learned that one of the adjacent property owners was a veterinarian. My attitude is that life is tough enough without having to worry about your neighbor's property being an incubator of exotic barnyard diseases. We called it quits for the day and made no commitment to try again at a later date.

We returned to being tourists the next morning when our sons, Brian and Scott, drove up from San Francisco. Scott was living there at the time, and Brian was visiting him that weekend from Pacific Palisades in Southern California. Both had been to the valley before and were intrigued with their parents' plans to buy real estate. At the imposing and baronial Hess Collection, a magnificent 210-acre winery, we sipped wine, browsed in the gift shop, and toured the extensive art collection that makes it one of the most fashionable stops in the valley. Located in the hills west of the city of Napa, it was originally built in the 1930s by the Christian brothers and was operated as the Mont La Salle Winery until the 1980s when Swiss entrepreneur Donald Hess leased it.

Our next stop was Saddleback Cellars in Oakville. It's a winery housed in a utilitarian cinder-block building on the floor of the valley and offers none of the frills we had just seen. Its appeal was the opportunity to sample blue-ribbon Zinfandel and Cabernet Sauvignon poured by owner and winemaker Nils Venge. Nils holds the distinction of being the winemaker of the Groth Winery's 1985 Reserve Cabernet Sauvignon, the first California wine given a 100 rating by world-renowned wine critic Robert Parker. A rugged outdoorsman of Danish descent who could have appeared in Marlboro ads, Nils stood behind the barrel that functioned as his serving bar and chatted amiably about the wines he made and the challenges of being in the winery business. I had met him years earlier and enjoyed his unpretentious approach to winemaking and life in general. Like the Paines, he sensed that the wine business was on the upswing, and the outlook was encouraging.

Midafternoon found us in St. Helena, where the banner across Main Street heralded the Harvest Festival, an effort by local merchants to sell their end-of-the-season inventory to the torrent of visitors that had flooded into the valley. My family wandered through the booths and displays that brimmed with an eclectic selection of food, books, crafts, clothing, and souvenirs while I took a real estate brochure from a sidewalk dispenser and found a shady bench on Adams Street. The listings covered the gamut from starter homes in Napa to large parcels in Calistoga. Tourists and locals passed by as I pondered the options, and eventually one of them stopped.

"Can I help you?" inquired a sandy-haired man dressed in chinos and a plaid shirt, who looked to be in his mid-forties.

"Not really. I'm just looking."

"I'm in real estate," he said. "All the people who ever bought from me were just lookers at one time."

"True enough. But I may never be a buyer."

"But then again, you may be." He smiled and his eyes twinkled as if to say, "If you expect to get rid of me, you're going to have to do better than that." His manner was persistent but not obnoxious.

"Let me introduce myself," he continued. "I'm Steve Ericson. I'm a realtor and have lived in the valley almost all my life. Maybe I can help you."

Reading listings was going to get me only so far. I needed more information, and Steve's interruption seemed like an opportunity to get some. I explained my interest in vague and imprecise terms while Steve listened and nodded at appropriate times. As happens so often, trying to explain something served to sort out and organize my thoughts. Steve had become like a real estate shrink, and he stepped into the conversation when I finally ran out of words.

"I think I understand," he volunteered. "And if it makes any difference, you're not the first person to start a search here not knowing what you want. Most people don't understand what the valley offers, so they aren't sure what to look for."

It might have been a response crafted in a realtor's training class, but coming from Steve, it didn't seem slick. If true, it was reassuring to know that confusion and uncertainty were not uncommon. Steve waded into the subject and it was immediately apparent that he was very knowledgeable and well informed about the valley. The conversation continued when Patricia and the boys returned. In the lengthening shadows of late afternoon, I explained to Steve that we were leaving in the morning and that I would not be back until February. In the meantime, we agreed to stay in touch.

Chapter 3:
Getting Started

My attraction to the West, and California in particular, was not new. It had been simmering for almost forty years, ignited by a trip after my junior year in college. Two friends and I drove a blue Ford station wagon across the country. We used Route 40 as our main thoroughfare, since the interstate system was still under construction. I had grown up in north Jersey and had never been further west than Pennsylvania, so seeing magnificent Rocky Mountain landscapes, bold Sierra Nevada terrains, and rugged Pacific shores was a new experience. I repeated the journey two years later, and in the years that followed the West was my destination as often as time and money permitted.

How much of its appeal was based on tangible differences with the East Coast and how much was in my imagination has never been clear. It didn't matter. It was a psychic treat just going there. The West sprawls across the landscape, offering beautiful vistas, open spaces, and big sky. It is less crowded and, while development certainly has increased over the years, it has maintained a fresh feel. The people there seem more independent and self-reliant than in other parts of the country, and they possess a twenty-first-century version of the pioneer spirit and heartiness that inspired past generations to stretch the country from coast to coast.

My fascination with the West is not unique. Lewis and Clark were the first to cross the continent, beginning what historian Stephen Ambrose called the "quintessential American act of heading west" (*USA Today*, November 4, 1997). Even lifelong New Englander Henry David Thoreau, who spent two introspective years in a cabin at Walden Pond near Concord, Massachusetts, once wrote, "Eastward I go only by force; but westward I go free." I was a victim of my American genes.

Patricia and I had moved to Anaheim in the mid-1960s and visited the Napa Valley for the first time in 1967. We liked it enough to return the following year. A few months later we were transferred to Ohio, and a few years after that Maryland became our home. Trips to Southern California, Nevada, Colorado, Arizona, and New Mexico stoked our interest in the West, but the strongest pull always came from the Napa Valley. Every visit there was accompanied by the desire to own land, but our priority in those days was raising a family, not investing in real estate. Now, however, our three children were grown and we were empty nesters. Susan, our oldest, had gone into banking after graduating from business school and was living in New York. Brian was a hotel developer with a small Los Angeles company, and Scott was attending business school in Palo Alto, California.

We were on the threshold of a lifestyle change. Retirement was two or three years away, and it was time to decide what the next phase of our life should be. The conventional wisdom that people who retire should not make drastic changes in their lifestyle sounded wise to us. Drastic changes weren't on our agenda, but we wanted to make some interesting ones.

Neither of us was ready for a sedentary life. A few years shy of sixty and enjoying good health, retirement would open the opportunity for new adventures. And in some yet-to-be-defined way, the Napa Valley fit those plans. My goal was to buy some "dirt" there, but Patricia was noncommittal. Perhaps, I thought, the right location would sway her.

That was the challenge: finding the right location. Prospective buyers quickly learn that they have to unscramble the valley's mystique and locate a home, vineyard, or empty parcel from a wide-ranging selection of choices. They can be "up" valley (generally defined as St. Helena and north) or "down" valley, which I defined as south of Zinfandel Lane. They can be in one of the few incorporated towns or on county land. They can buy a house. Or a vineyard. Or a house with a vineyard. Or a house at one location and a vineyard at another. Or just bare land on which to build or plant grapes. They can have a hillside location or one on the valley floor. They can have morning sun or afternoon sun. The combinations and permutations have to be understood, sorted out, and prioritized. It makes no sense to spend time hunting unless you know what your quarry looks like.

We had moved often enough that the prospects of a search were not intimidating to us. In fact, this one posed less pressure than others because there was no deadline. The big difference was that each parcel in the valley has unique characteristics, such as the amount of land, the type and number of structures, and the surroundings. Since it was not likely to find two similar pieces of property, our search would be different from others we had made in places with relatively homogeneous suburban neighborhoods and developments. Nothing we saw in the Napa Valley was likely to have something similar across the street, down the road, or, for that matter, anywhere else.

Refining our selection criteria was going to be essential. At the top of the list was a house with a vineyard. After that was a house without a vineyard but with the possibility of buying one nearby. A well-situated parcel for building a house and planting a vineyard was next. A vineyard all by itself held no interest. A hillside location had slightly more appeal than the valley floor. Being isolated was out. Being able to see other homes was important, but there was no need to be right on top of them. Finally, the area having the greatest attraction was bounded by Zinfandel Lane on the south and Lodi Lane on the north.

Patricia and I agreed that I would take the lead and screen the possibilities, since the desire to buy property was mine. When I found something worth considering, Patricia and I would look at it together and make a decision. Flying back and forth across the country was enough of an expense and inconvenience to serve as a check on my enthusiasm and a curb on the number of trips I would make. We were ready to launch.

A business trip to San Francisco in February 1996 provided the opportunity to spend a weekend in the valley as a potential real estate buyer rather than a tourist. The morning was sunny and the air was crisp. The billowy clouds hanging against a deep-blue sky looked like the handiwork of a landscape artist. The leafless vines were barren and stark, providing an unrestricted view of how vines, stakes, and trellis wire create the vineyard infrastructure. Their stark appearance contrasted with the green groundcover and yellow mustard flowers at their feet.

Steve was waiting in his office on Main Street in St. Helena, not far from where we met. We had spoken a few times by phone, and he had lined up some properties to visit. It was time to discover what real estate gems the valley had available for me.

"Let's look at these," he suggested, handing over a stack of multiple listing sheets. "As you begin to see what the valley has to offer, you'll get a handle on what will satisfy you and Patricia."

"I won't be able to tell that much just by reading them," I said, "so let's start looking."

Steve, it turned out, was as interested in getting acquainted as he was in showing real estate. He wanted to know what made me want to be in the Napa Valley. I told him about my first visit in 1967 and some of the later ones, volunteering that we were not considering any other places. After looking at two listings, we pulled off the highway in front of the parcel two miles north of St. Helena that I had seen the prior September. It was still on the market at the same price. Steve described it briefly and then sat with his elbows resting on the steering wheel as he listened to my reaction. He got the whole story, including my dislike for living next to a veterinarian.

He cocked his head and looked a bit quizzical. "That's amazing. I've been in this business a long time and have heard a lot of things, but never something like that."

"It probably comes from growing up in the suburbs," I volunteered. "Being in the country will take an adjustment. We know we have to make tradeoffs, but having sick animals as neighbors is not one I'm willing to make."

"This should be interesting," he remarked with a grin, and he turned on the ignition to head to our next destination.

"There's a house I want to show you that's been on the market awhile. It's the best-built house I know of and has lots of space. Sixty-three-hundred square feet. Right now it's owned by the Bank of America."

He turned onto Byrd Hill Road and followed it up from the valley floor. The road cut diagonally in front of a charming Victorian home and continued on what had become a steep, forested slope. Above us and slightly to the left, a large house had been built into the side of the hill, sharing the terrain with towering pine trees. The road curved back across the slope below the big house and passed in front of a smaller one. Steve explained that it was a parcel of eighteen and a half acres with two houses.

The road got steep again as we approached the big house. Its exterior was toast-colored cedar siding. Once inside, it became apparent that the house was built with quality in mind. One large room with rich reddish-brown floors served as a combination living room and family room. A large stone fireplace dominated the east wall. The kitchen was large and the appliances were suitable for a small restaurant. Three other rooms on the first floor could have been used for an office and bedrooms, and the lower level could accommodate three or four more rooms and a wine cellar.

The elevation of the large deck on the south side of the house provided a good view of the valley. The eastern slope could be seen through an opening in the tree canopy, and below on the valley floor was a French chateau-like house and vineyards. This place was head and shoulders above anything I had seen, but its size was a bit overpowering and I didn't see a place for a vineyard. On the positive side, it was owned by a bank that had reduced the price 20 percent since listing it on the market a year and a half earlier.

"The smaller house can be rented to give you some income," Steve observed as we drove back to it. It was older but comfortable, and it had a small pool on the south side. The views from the porch were even better than those from the larger house. The idea of having two houses and renting one was a new consideration that required some thought.

Steve guided me through three other places as we continued the chatter of getting to know one another. He wore well. My initial impression that we could work together seemed to be holding. By the end of the weekend, I had begun to develop a feel for availability and prices. He was willing to show me as much of the valley as I wanted to see and even remarked over lunch that he had worked with one couple for two years before they found something. I winced at the thought of a two-year ordeal, but it was reassuring to know that my realtor had that kind of staying power.

Another trip a few weeks later gave me the opportunity to look at two parcels of land that had neither a house nor a vineyard. Their appeal was "killer optics." It was a term Steve used sparingly, and it meant the surroundings and the views were exceptional.

A sixty-acre parcel located a mile up the Oakville Grade from Route 29 was almost at the valley's midpoint. High above the valley floor, it provided a perspective to the checkerboard of roads and vineyards in springtime colors that emphasized the agricultural nature of the region. The property itself was steep and rolling and offered two or three good building sites, but there was little possibility for a vineyard. Its big shortcoming, however, was remoteness from neighbors and civilization.

A forty-acre parcel on Spring Mountain Road was only five minutes from downtown St. Helena. It, too, was rolling but less steep. The owners were in the process of planting a vineyard and had staked an ideal setting for a house. The view south revealed the valley widening as it approached San Pablo Bay. A portion of forest on the mountains to the west had been shaved away to make room for a meticulous four-acre vineyard. It was promising. Our next three stops were houses with small vineyards that had no appeal, and they were quickly eliminated from consideration.

My positive reaction to the Spring Mountain parcel prompted Steve to caution me not to wait too long to make up my mind. He pointed out that the market was beginning to come alive. It would be out of character for a realtor to ever tell a prospective buyer that the market was cold and there was no pressure to make a decision, so I listened to his advice and nodded but made a slight detour on my way to the San Francisco Airport to see the parcel again. The views were exceptional and the price was near what we were willing to pay. It might be worth having Patricia take a look.

On the phone with Steve two weeks later, I mentioned how much I liked the views at Spring Mountain. "Take it off your list," Steve volunteered. "That's in escrow already. It closes later this month."

My feeling of surprise was replaced almost immediately by one of disappointment. I had seen about twenty properties by that time and it was the one I liked the most. Somehow I had made the careless assumption that, despite Steve's admonition, it would remain on the market until I made up my mind.

"The market's starting to heat up," he remarked. Maybe that wasn't just hype.

Chapter 4:
Understanding the Climate

A forty-mile drive north and slightly east from San Francisco takes you to the southern end of the Napa Valley. Here it borders San Pablo Bay, which is the northwestern part of San Francisco Bay, and is five miles wide. The valley narrows as it runs for thirty-eight miles in a northwesterly direction from the city of Napa to Calistoga.

It is bounded on the west by the Mayacamas, a spine of mountains separating it from the Sonoma Valley. Densely forested for most of their length, the Mayacamas begin to flatten out at their southern end before disappearing into the marshy delta at the bay. A drive along Mount Veeder Road winds through them and takes you past redwood, Douglas fir, black oak, and live oak trees that thrive in cool, damp regions. With an elevation of 2,677 feet, Mount Veeder is the highest point in the southern part of the valley.

The boundary on the east is provided by the Vaca Mountains. They appear less brawny than the Mayacamas, and the sparse digger pine, scrub oak, and madrona trees at their southern end are evidence of less rainfall on the east side of the valley. While rock outcroppings and craggy surfaces also distinguish this end of the range, two-thirds of the way up the valley the difference in tree coverage between the two sides diminishes. Both the Vacas and the Mayacamas are part of the Coast Range system that parallels the coastline.

Sprinkled between Napa and Calistoga are the communities of Yountville, Oakville (unincorporated), Rutherford (unincorporated), and St. Helena. The main road up the valley is Route 29. As you drive it, there is little to indicate any elevation change until you pass the Veteran's Home turnoff in Yountville and find yourself on the route's only substantial incline. Napa is at sea level, while Calistoga has an elevation of 303 feet.

About four miles past Calistoga, the Mayacamas and Vacas come together to pinch off the valley and make it a large box canyon. It is here that Mount St. Helena is located, a peak 4,344 feet high that dominates the view north of Yountville. The uniqueness of the valley's soil and climate is what makes it so well suited for viticulture.

The valley was sculpted millions of years ago by a series of earthquakes and volcanic eruptions. The uplifting, faulting, and compressing of earth created the Vacas, the Mayacamas, and the knolls, slopes, and rolling areas that give the floor of the valley its character. Great rivers rushed out of the mountains, depositing disintegrated volcanic rock and silt on slopes and across the floor. As a result, the soils on the hillside are typically rocky, shallow, and less fertile that those on the floor. However, where they are mineral rich and well drained, vines flourish and impart an intense flavor to their fruit that produces outstanding wine.

The Napa River runs the length of the valley and helps channel the runoff from the rain. The navigability of its southern end made the city of Napa a port that contributed to its early prominence and aided its growth. The river can overflow its banks and cause widespread flooding during December, January, and February, when the valley gets slightly more than half of its annual rainfall. The flooding that occurs periodically caused the state to designate much of the valley floor a flood plain. Recently the county began a major project to widen the river in order to minimize future flooding.

"It would be difficult to construct an area more favorable to growing high-quality grapes than the one we have in Napa

Valley," says Jeff Popick, whose weather column appears in the weekly *St. Helena Star*. Popick is a transplant from Maryland who has had a lifelong interest in meteorology. His degree in viticulture from UC Davis led him to vineyard-management positions with two wineries before he developed his own Chameleon label. His weather column is a sideline.

"The warm, dry summers and cool, wet winters of the region are what geographers and meteorologists call a Mediterranean climate," Popick explained to me. "A dome of high pressure off the Pacific Coast in the summer forces storms to the north. The dry climate that is created protects vines from diseases that flourish where there is high humidity. Northwesterly winds bring cold water to the ocean surface that cools the air and creates the banks of fog associated with San Francisco. The fog and cool air that is funneled up the valley moderates its summer temperatures."

These conditions enable grapes to flourish during a growing season that starts in mid-March, when the average daytime temperature begins to remain above fifty degrees, and ends in October. Except for an occasional light shower, there usually is no rain between mid-May and late September.

The grapes remain green and hard until early August, when a phenomenon called veraison occurs. This is when red grapes begin to turn red and white grapes begin to turn a soft green or yellowish color. Sugar content increases until the grapes are mature and ready for harvest. By October, the valley begins to experience intermittent rainfall. The vines become dormant in the winter when they are pruned in preparation for the growth cycle to repeat itself.

The challenge for winegrowers is to select a grape variety best suited for their location and soil. Professor Albert Winkler of the University of California at Davis made a major breakthrough in the 1950s by identifying climatic zones based on the total amount of heat accumulated during a growing season, since grape varieties differ as to how much heat they need to ripen. Chardonnay and Pinot Noir prefer the region closest to San Pablo Bay that is most influenced by the cooling fog from San Francisco Bay. Just to the north is where Sauvignon Blanc and Merlot thrive and where Cabernet Sauvignon grows well. Still further north is where Cabernet Sauvignon, Zinfandel, and Sangiovese are most at home.

Sunlight and warm temperatures produce sugar in grapes, and sugar level, acid, and flavor are the most important factors in determining ripeness. Grapes are well suited for making wine because of their high sugar content, usually 20 to 26 percent. In addition, 70 to 80 percent of their weight at harvest is water. It is the process of fermentation that converts a grape's sugary water into wine. The formula for fermentation is:

$$\text{Sugar} + \text{Yeast} = \text{Alcohol} + \text{Carbon Dioxide}$$

While yeast occurs naturally on the grape skin, it is not reliable for modern winemaking, so wineries usually replace it with varieties that have been developed in laboratories. The fermentation process ends when the yeast has converted all the sugar into alcohol. A wine made from grapes that have a 24-percent sugar content when picked will have an alcohol percentage that is 56 percent of 24, or 13.4 percent.

Acid level gives wine its crispness and prevents it from tasting flabby. The daily-temperature swings of 35 to 40 degrees experienced during the growing season are excellent for producing sugar and for maintaining the desired levels of acid. Fortunately for vineyard owners and harvesters alike, grape varieties ripen and become ready for picking at different times.

The thirty-two different soil types that have been identified in the Napa Valley enable the area to grow more varieties of wine grapes than any area of comparable size in the world. Bordeaux is almost eight times larger than the Napa Valley, but the French government allows only certain grapes to be grown there. They include Cabernet Sauvignon, Merlot, Cabernet Franc, Malbec, Petit Verdot, Sauvignon Blanc, and Semillon.

More than sixty grape varieties are grown in the Napa Valley. The six most popular are grown on 90 percent of the planted acreage. In descending order of acres planted, they are Cabernet Sauvignon, Chardonnay, Merlot, Pinot Noir, Sauvignon Blanc, and Zinfandel.

Cabernet has been the most widely planted variety since 1992 when it edged past Chardonnay, reflecting a shift toward full-bodied red wines among American consumers. It also attracts the highest average price per ton of the six most popular varieties.

The vineyard acreage in the Napa Valley represents approximately 7.2 percent of California's vineyard acreage, but produces approximately 3.9 percent of its wine-grape tonnage. It is the drive for grape quality that accounts for lower yields than in places like the Central Valley and establishes Napa as the premier wine-growing region in the state.

Chapter 5:
"There's No Quit in That Guy"

Steve was not one to let any grass grow under our search. He called every few weeks to describe the new properties that had come on the market. On one occasion, I chided him for including one that was way off target. He laughed and regaled me with the story of a client who had been absolutely insistent on what he wanted right up to the day he bought something entirely different. I couldn't fault his approach.

He particularly enjoyed showing me anything "heroic" that appeared in the listings. It was his term for parcels in out-of-the-way places that were accessible only by narrow, winding roads over rough terrain. A combination of my curiosity and Steve's spirit of adventure resulted in us jostling across, through, and over more hillsides, dry creek beds, and gorges than I ever expected to find. Anything he described as "heroic" was bound to be memorable, and he had one lined up when Patricia and I arrived in April 1996.

The Bartlett property was seventy acres on the side of a hill. It had neither a house nor a vineyard, but was described as having excellent views and good soils. Steve estimated that at least ten to twelve acres could be planted. Our route took us from the paved portion of Crystal Springs Road to a rugged dirt trail. Steve clung to the steering wheel, and our seatbelts kept us from bouncing off the roof as we crossed a shallow stream and switch-backed our way up a tree-covered slope. After half a mile, he stopped in a beautiful meadow adorned with green grasses, brown weeds, wildflowers, and a sprawling oak tree. We were a few hundred feet above the valley floor at a point where the valley is about a mile and a half wide. Across to the west was a forested hillside interspersed with vineyards. Mount St. Helena anchored the view to the north, and hillsides, vineyards, and trees shared the terrain to the south.

"Nice, isn't it?" Steve remarked as Patricia and I got out of the van. Two hawks with broad wingspans rode the thermals below us in a scene of sublime quiet. It was a perspective of the valley we had never experienced.

"Absolutely," I responded, knowing my body language had given away the answer.

It had remarkable views, great potential, and a reasonable price, but it was remote, the access was rugged, and we would be starting from scratch. Despite all the pluses, Patricia's enthusiasm was much less than mine, and we decided to scratch it from our list. I was disappointed, because it was so reminiscent of the parcel on Spring Mountain Road that had slipped through our fingers.

Two months later, Patricia took a solo trip to the Bay area to visit Scott, and she looked at the Bartlett property again. She warmed to it, and during a telephone conversation we decided to make an offer. Steve prepared the paperwork and dropped it off with the listing agent early the next morning while Patricia boarded a flight to return to Maryland. He called me with an update while she was still in the air.

When she arrived home, her first words were, "Have you heard from Steve?"

"I have. You were there the day of the Napa Valley Wine Auction. One of the attendees had looked at that parcel twice before and dropped off an offer the same day you did. He offered almost the full asking price with no contingencies. Steve wondered if we wanted to increase our offer or withdraw our contingencies. I told him we had no interest in doing that and said we'd just keep looking."

The rising economy was bringing buyers back into the Napa Valley real estate market. The period of declining values the Paines had described seemed to be fading into history. We were fortunate to have a bird dog like Steve in our corner. He had his finger on the valley's pulse and was always on the lookout for new possibilities. His natural optimism and positive outlook prevented me from feeling disappointed. I didn't know how many trips it would take to find our place, but I was mentally digging in for the long haul.

It seemed like an appropriate time to make an all-out assault on new listings, so Steve lined up ten properties for my next trip in July 1996. The first was a sixty-acre parcel that was part of the Marston Ranch in St. Helena. The approach to the property was through a stand of large trees that not even the midday sun could penetrate. We scratched it.

A thirty-three-acre parcel on Taplin Road, which was not yet on the market, was appealing. The owners were rumored to be splitting up, and there was speculation that they would be putting it on the market. My interest in it gave Steve another reference point for evaluating properties for my consideration.

We spent the day checking out each property on his list. Unfortunately, I gave all of them a thumbs down. The notes I made of each visit kept them from blurring into an indistinguishable mass of real estate.

In mid-afternoon, we turned off the Silverado Trail into the Prinn Vineyard located behind the Sterling Winery. Here was a forty-two-acre parcel with twenty-five acres planted in Cabernet Sauvignon, in addition to a reservoir and a home site on a knoll overlooking the vineyard. We toured it and parked to discuss it when a truck pulled up next to us. The driver rolled down his window and Steve explained who we were.

"I'm Jess," explained the newcomer. "I manage this place," he said, pointing toward the vineyard. "What do you think?"

"The vines don't look very healthy," I replied.

He nodded and volunteered, "They're diseased. Phylloxera. They've got to be replanted."

"What will that cost?"

Jess gave us an estimate of the cost of ripping out and replanting the vineyard. He explained that the first harvest of about a ton per acre would be in the third year. Production would likely top out at five tons an acre in the fifth or sixth year and continue at that level for ten or fifteen years. Maybe longer. Since he managed about three hundred acres in the valley, his information was based on experience.

"That was worthwhile," I volunteered to Steve after Jess drove off.

Steve agreed and watched me make some basic calculations on a pad regarding the cost of tearing out the old vineyard and replanting. This was a big parcel, and the overall cost was once again outside of my comfort range. When Steve ribbed me for being too conservative, I reminded him that being almost sixty and about to retire was not a time to make a financial blunder.

Knowing I had to fly home early the next morning, he dropped me off back at the comfortable surroundings of the Wine Country Inn. "When you review your notes, don't forget the big house on Byrd Hill. You looked at it on your first trip back in February. It's the best value around."

"I won't forget it because I know you won't let me," I joked.

None of the properties wowed me, but the Prinn Vineyard deserved some further consideration and Byrd Hill was still on my mind. It was the best value on the market and was looking better and better with prices trending upward. The most interesting part of the day, however, was getting to know Steve better.

The accident that put Steve into a wheelchair had occurred fourteen years earlier. He had been working on a construction site and had fallen from ceiling joists, injuring his spine and causing him to lose the use of his legs. His body may have been injured, but his soul and character were undamaged. After the accident, he mastered getting from his wheelchair into his van and reversing the process at his destination. He never complained or whined and rarely mentioned his limitations. He was playing the cards that life dealt him. No more, no less.

Steve and his wife Marla were the parents of an eight-year-old son and a two-year-old daughter when I met them. His daughter from an earlier marriage was an architectural student at Cal Poly in San Luis Obispo, and his mother Greta had been the first woman mayor of St. Helena and now was also a real estate agent. Above all else, Steve was a natural-born optimist. If I didn't like something about a house we looked at, he had two or three suggestions for dealing with it. If I didn't like something about a parcel of land, he suggested how it could be re-graded or altered in some fashion. There was nothing that couldn't be fixed, except, of course, location.

His grittiness, strength of character, and persistence had made the results of his accident irrelevant. He was never down and had the knack of finding the positive feature in anything. One of his longtime friends commented to me one day, "There's no quit in that guy." It was a remark that fit Steve to a tee. His benefit to a prospective buyer was his knowledge about the valley and the people who lived there—and his patience. He never hurried or showed any annoyance with my uncertainty as we went from property to property. We were compatible and I was learning about the valley day by day and week by week. Being in Steve's company was an education.

Chapter 6:
Creating an Agricultural Preserve

Acre after acre of chest-high vines and trellises cover much of the Napa Valley floor and produce an appearance of orderliness. Following the contours of the land, they create geometric patterns that stretch in all directions. Masked by this visual harmony is the intense bickering between vineyard owners and wineries, old-timers and newcomers, politicians and citizens, and among families that accompanied the 1960s debate over land use in Napa County. Their discussions followed a bumpy and contentious path but eventually resulted in a simple statement of purpose that "agriculture is and should continue to be the predominant land use." It was a decision that gave testimony to the valley's unique beauty and its winemaking heritage. It is implemented by two ordinances. One, called the Agricultural (or Ag) Preserve, regulates land use on the valley floor, while the other, called the Agricultural Watershed Ordinance or the Hillside Ordinance, covers the surrounding slopes.

Their creation was an outgrowth of our nation's mushrooming population during the twentieth century. Urban sprawl produced suburbs and created the relentless move of civilization into rural areas. Zoning and land use were hotly debated issues, and friction became intense between developers and environmentalists. The role of government began to expand and in the process the wetlands, the spotted owl, and redwood preservation became topics that captured national headlines.

California played a marquee role in this story, since its rate of growth was multiples greater than the national average. While the number of people in the United States doubled between 1900 and 1950, the population of California increased seven times, rising from 1,485,000 to 10,586,000. Between 1950 and 1960, California's 48-percent population growth was almost four times greater than that of the rest of the country.

Virtually every county and community in the Golden State felt the pressure caused by this human explosion. New homes were built at a dizzying pace. Freeways proliferated. Orange and Santa Clara Counties, feeling the squeeze from Los Angeles and San Francisco respectively, were beginning to lose their agricultural character. Farms, orchards, and vineyards were gradually giving way to streets, driveways, backyards, and shopping centers. It was against this backdrop that Napa County began to contemplate its future.

California had developed an extensive freeway-construction plan by the 1960s that included the expansion of Route 29 into a divided, four-lane road up the center of the valley. This plan, approved by the County Board of Supervisors, put the region on the threshold of a huge transformation. Then some members of the board began to express reservations about aggressive development as they and their constituents watched the widening of the southern end of the highway in Napa. The realization set in that this broad asphalt strip and its accompanying interchanges would be a huge scar on the landscape.

Citizen groups opposed to the freeway were forming by 1965, one of them under the guidance of Dorothy Erskine, who lived in San Francisco but also had a home in the Napa Valley. Over the years, she had become one of the Bay Area's most ardent champions of open space. She vocally and energetically opposed the spread of concrete and asphalt and spent countless hours promoting, cajoling, and preaching the benefits of natural beauty throughout the region.

One spring evening, she held a meeting at her home just outside of Calistoga. It was attended by vineyard and winery owners, bankers, farmers, lawyers, and shopkeepers. The large number of guests made it tight and warm inside, and they sipped white wine and munched on cheese and crackers until Dorothy called the meeting to order. She knew most of the attendees and they knew each other. She explained her interest in helping protect the agricultural heritage of the valley that had served all of them so well. Then she listened as they expressed their hopes, dreams, and concerns about the future.

Among the attendees were Jack and Jamie Davies, newcomers to the valley who had recently purchased the Schramsberg Winery on Diamond Mountain, midway between St. Helena and Calistoga. They didn't know Dorothy but had learned of the meeting from friends. The Davieses were transplants from Los Angeles, where they had observed the sprawl that had engulfed Southern California. From their travels in the state, they also were aware of the wave of real estate development that was transforming the counties south of San Francisco. They understood the vulnerability of the Napa Valley better than most people. Jack's career in business had made him an excellent problem-solver, and he plunged into the discussion. "Each of us knows what we want," he said. "But to accomplish anything worthwhile, we need a plan for working together. We'll only succeed if we can focus our energy and effort toward a common goal." His remark was met with a sea of nodding heads.

"Dorothy picked Jack to be the Citizens Committee chairman at the meeting at her house," Jamie Davies explained to me many years later in the warm, picturesque reception room of her winery one rainy winter afternoon. Jamie was a small, almost diminutive woman who would have appeared fragile but for the resolve in her eyes and the firmness of her words. "Jack had leadership skills from his business career and had the ability to bring people together and form coalitions. And it was apparent that we needed a coalition of citizens. Dorothy was dedicated to conservation and preservation. She was an effective spokesperson and was good at getting together people who cared about the Napa Valley. You can't point to any one person and say he or she started it, except maybe Dorothy. If she didn't start it, she was at least the catalyst."

What Dorothy helped create was a wave of community interest to combat the forces favoring development of the valley. People began to express the feeling that the Napa Valley should be considered a national treasure, and they wondered what steps had to be taken for agriculture to be accepted as the best use of the land. Unless this occurred, the valley ran the risk of becoming like Santa Clara and Cucamonga, where many vineyards had been turned into strip malls, parking lots, and housing developments.

Mental pictures like that helped galvanize the community. The new Board of Supervisors wanted to preserve the agricultural heritage of the valley but needed widespread support. They realized it was a stormy issue that could, and did, divide families and splinter friendships.

Vintners and farmers were active on both sides of the argument, getting petitions signed and drumming up support. Louis Martini considered it a wonderful idea and worked energetically in its behalf, while John Daniel of Inglenook regarded it as arbitrary and unconstitutional and opposed it with fervor. With lifelong friends reluctant to antagonize one another, it was better for a newcomer like Jack Davies to take the lead, and he agreed to do so.

"I remember one critical day when things were tense and we were concerned whether we were making progress," Jamie reminisced. Their committee had arrived at a meeting with the Board of Supervisors with a huge pile of letters from people who couldn't attend but who supported the Ag Preserve. Since rooms in the county office building and the courthouse were not big enough to handle the crowds that gathered, it was necessary to use the auditorium of a high school in Napa. "The board was influenced by the letters," Jamie explained. "But the real difference was made by Eugene Trefethen."

Eugene had held the positions of president and vice-chairman of Kaiser Industries. He was the energetic local leader of a division of a company that had built a formidable reputation for tackling big jobs, like building Liberty ships during World War II. He was one of the most respected business figures in the Bay Area. He was looking to buy property in Napa County and had developed an interest in the old Eschol Winery on Oak Knoll Lane.

His remarks to the supervisors made it clear that he would buy land here only if the Ag Preserve was created. He was looking for long-term use of the land as a vineyard and didn't want to worry about housing and urban sprawl. That got the attention of a lot of Napa folks. Here was a captain of industry saying that Napa offered something he cherished—land for its agricultural value rather than its development potential. People listened to him. What he said and how he said it had a profound influence on the outcome.

Another person who watched the change evolve and helped shepherd it was Jim Hickey. Jim, a native of Michigan, had been stationed in the Bay Area during World War II before being shipped out to the Pacific. Like so many other GIs of his era, he had resolved to return to California if the opportunity arose. He was working as the planning director of Stark County, Ohio, in the early 1960s when he learned of the newly created position of executive director of the Association of Bay Area Governments. He threw his hat into the ring and after a successful round of interviews was selected to fill the post in March 1964. Jim watched the developments in Napa from this new vantage point and then jumped at the chance to become planning director of Napa County.

Jim, now retired, explained that the people opposed to the change wanted to be able to sell off a few acres when they needed cash. Farmers and winemakers frequently experience financial ups and downs, and land sales help them get through tough times. Under the proposed Ag Preserve, the county would go from a one-acre minimum parcel size to a twenty-acre minimum. That meant a person would need at least forty acres to subdivide his or her property. No parcel smaller than twenty acres could be created, which meant selling off a few acres no longer would be possible.

Jim drafted the ordinance and became a target of criticism, but his commitment never wavered. After almost a dozen public hearings, much debate, and rancorous dialogue, the planning commission adopted the Ag-Preserve measure unanimously on January 15, 1968. The Board of Supervisors did the same on April 9. The goal was spelled out in the first section of the ordinance:

> *This district classification is intended to be applied in the fertile valley and foothill areas of Napa County in which agriculture is and should continue to be the predominant land use, where uses incompatible to agriculture should be precluded and where development of urban type uses would be detrimental to the continuance of agriculture and the maintenance of open spaces which are economic and aesthetic attributes and assets of the County of Napa.*

After the Ag Preserve passed, wineries knew they could invest and have their investments protected. They didn't have to worry about a housing development coming in next to them and eventually forcing them out of business with their complaints. There were about sixty-four wineries then, and there are more than four hundred today.

"If the Ag Preserve had not been created, Napa would be indistinguishable from the counties around us," observed Jamie Davies. "You wouldn't have the sense of being somewhere with character and natural beauty—a place where people, agriculture, commerce, and the environment are in balance."

"You have only to look at neighboring counties to see what we would be like without the Ag Preserve," Jim pointed out. "Keep in mind that some of the appeal of Napa isn't what you see but what you don't see. The small signs that identify wineries and designate appellations are an example. They are tasteful and unobtrusive because of our regulations. They're part of the environment we wanted to create and still need to protect."

While the Ag Preserve has widespread community support today, it still rankles some people as an example of government interfering with individual property rights. Those opposed to government regulation might see it as the worst thing that ever happened in the county. But those interested in preserving the area for something other than subdivisions are likely to see it as a unique coming together of people in the community to shape the future constructively. Fortunately it has turned out well.

The Ag Preserve's protection of agriculture provides the blueprint for land use on the valley floor outside the five incorporated communities of American Canyon, Napa, Yountville, St. Helena, and Calistoga. Subdivisions are nearly nonexistent, and each parcel possesses its own individuality. The minimum parcel size of 20 acres created in 1968 was increased to 40 acres in 1973 and later to 160 acres. Smaller parcels that existed prior to each change were "grandfathered" and cannot be subdivided. The result is a finite supply of parcels, rather than the constantly expanding supply found in most communities. Measure J, a ballot measure approved in 1990, gives citizens the right to decide if major changes are to be made in areas zoned for agriculture until 2020. Measure P, approved by voters in 2008, extends this to 2058, thereby protecting the Ag Preserve concept well into the future.

Citizen activism was the essential ingredient in confronting the contentious problem of land use. Resolution was reached through debate and discussion. The result stands as a testament to the foresight and determination of a handful of partisans who rallied their neighbors during the 1960s to create a unique environment that has won the praise of twenty-first-century Napa residents and visitors and the envy of its neighbors.

Chapter 7:
Two Big Decisions

A decision had to be made about the house on Byrd Hill. It didn't exactly fit our specs, but it was the best value on the market. Steve arranged for Patricia and me to tour it with the builder, Kurt Becker, in September 1996. His pride was evident as he pointed out the features and workmanship that gave it such appeal. It was wired for ten phones; the siding and decks were redwood; the floors were Brazilian cherry; and the railing, downspouts, and gutters were copper. The kitchen could have been in the Culinary Institute. Kurt also pointed out that a slope on the south side had the potential for a small vineyard.

"It's not what I had in mind," Patricia remarked when Kurt left. "But it's the best house we've been in."

"And the market is heading higher," I added. "Prices are up."

After discussing some of the work that would have to be done to make it suit us, we decided to make an offer. But by the time the bank's counteroffer arrived three days later, Patricia's enthusiasm had begun to fade.

"There's only enough space for a small vineyard," she pointed out. "Maybe an acre. That's smaller than we wanted. This isn't looking as good to me anymore."

"But it gets us started," I responded. "And rent from the smaller house would defray our expenses."

We realized we both needed to agree if this was going to work. Having an expensive piece of real estate 2,500 miles from home that needed time, money, and attention could easily become a marital nightmare if only one of us supported the idea. Patricia's interest in Byrd Hill had evaporated, while I saw it as a way to buy at an attractive price, ride the market up, and swap it for something better when we found it. Our points of view had collided and now she was firmly opposed to owning Byrd Hill. The deal was off. I explained our decision to withdraw our offer to Steve, who took it very philosophically.

"It's the best buy out there. But if you don't want it, it's not the right place for you. I'll tell the listing agent. She'll be surprised." He paused before asking, "Is your search dead or are you still going to look?"

"I'll still look," I answered. "I'm just not sure when I'll be back."

A few days later, Steve's mother, Greta, showed Byrd Hill to a couple who lived in the valley and wanted a bigger house. They loved it and made an offer below ours that the bank immediately accepted.

We heard nothing from Steve for almost four months, until he called in January 1997 to say the thirty-three acre parcel on Taplin Road I had seen in August was going on the market. The owners had decided to divorce and wanted to sell it.

Nestled in a bowl of land a quarter of a mile past the entrance to the Phelps Winery, it was bordered by property owned by vintner Joe Heitz. Most of the land was on a west-facing hill dotted with oak trees. The buildings – a very modest house, a barn, and an art studio used by the wife to do her painting – were located in a flat area at the base of the slope. At the top of the hill were another five flat acres that had been cleared of shrubbery, brush, and trees but were not being used. Power lines above this portion of the property were the only negative feature.

It had not yet been listed so the asking price was not known. Patricia and I looked at it and decided we liked it well enough to make an offer, knowing our price would set the floor. Steve got no response for almost three weeks until the owners responded with their listing price. It was much higher than we had expected. Trying to decide what to do next, I spoke to Joe Ryan, a business associate who always had a knack for peeling away irrelevancies and getting to the nub of an issue. Sitting behind his desk, he listened to my description of the situation, then leaned back in his chair with his hands resting on his chest, his fingers interlaced.

"Now if I understand it, you are looking to have a vineyard business and a place to spend maybe two or three months a year."

I nodded.

"Is it safe to assume your down payment would be about 20 percent of the purchase price?"

I nodded again.

"How many years would it take you to spend that amount if you forgot about the vineyard and only rented a condo or stayed in a hotel in the valley for two to three months each year?" he asked rhetorically. "From the numbers you gave me, I estimate you could do it for at least ten years. And you wouldn't have a mortgage payment. If you hated it after three or four years, you could walk away. No harm. No foul."

He was right, but his analysis didn't dampen my interest. That's when I realized I was not looking at this in economic terms. I was pursuing a dream, and the logical approach that had guided virtually every major decision in my life was being diluted by more emotional considerations. That was a sobering realization. People can make big mistakes when they wander into uncharted waters.

Dream or not, we couldn't consider making an offer anywhere near the asking price without feeling we had lost our minds. Ten days after the property was listed, a buyer stepped forward with a full-price offer and another appeared with a backup offer at the full price. Somewhere in the back of my mind, I was hearing "Strike Three!"

Chapter 8:
"Querencia" Explains It

Our disappointment over the sale of the Taplin Road property did not last long. We could not have spent that much money and thought we were doing the smart thing. It fit someone else's objectives better than ours. At it for slightly over a year, we had invested too much time, energy, and emotion in our search to throw in the towel now.

From each trip and contact with Steve, we learned more about the valley's geography, real estate, and weather and became acquainted with more people. We considered renting a home for a month or two, but decided against it. It wouldn't give us any better feel for the area than if we stayed at a hotel. We were serious about buying and didn't want to compromise with that position.

One challenge we faced was keeping our expectations in line with reality. We wanted to avoid romanticizing too much about life in the valley. Our lifestyle, habits, and interests reflected our upbringing and experience and were not those common to agricultural areas. Patricia had grown up in Cambridge, Massachusetts, and I was raised in Teaneck, New Jersey. Suburban living was in our veins. We realized that life in an agricultural preserve would be different, but didn't know for sure what that meant because we had not yet experienced it. Yet we expected to embrace it.

We had gotten accustomed to the idea that we would have a septic system and a well, and both would be our responsibility. Throughout our lives, utility companies had typically taken care of any water and sewer problems we experienced. If something went wrong, a call to customer service had produced a solution. It had always fallen to someone else to get those things fixed, but in our new environment it would be our responsibility. Not having lived with a well and septic before, we didn't know how much upkeep they would require or who we could rely on for help when it was needed.

Another adjustment to be made concerned the conveniences that had become a standard part of our lives. Living where spraying, plowing, and dust are commonplace was going to be much different than living where hiring a lawn service and pulling weeds were the closest we got to agriculture. In addition, our Washington, D.C. area home was a fifteen-minute drive from two shopping centers that offered as many goods and services as anyone would need in a lifetime. The museums and cultural activities of Washington, D.C. were just twenty-five minutes away. Nothing in the valley was comparable or as convenient.

The more we visited the valley, the more we understood it. The more we understood it, the more we liked the prospects of the change. Enough people had figured out how to live with septic tanks and wells that we could, too. A house doesn't need a large family room and a cavernous basement to be comfortable. And not having Starbucks, Macy's, the Gap, and Barnes & Noble a few minutes away would be a blessing, not a hardship.

Doubts about Patricia's interest in this venture had been resolved when she had agreed to make the offer on the Bartlett property. We had been at odds over Byrd Hill, but our joint interest in the Taplin Road property had put us back on track. During more than thirty years of marriage, we had both learned the fine art of accommodation and were practicing it as we went along.

Unlike Robert Mondavi, who recounts in his book *Harvest of Joy* the Italian family practice of giving glasses of water-diluted wine to children at dinner, I had no cultural or ancestral ties to wine. In fact, my first experience with wine had been disappointing. My father had been encouraged by his doctor to drink alcohol before dinner to dilate his arteries and relieve some of the discomfort caused by his heart condition. His original choice for a pick-me-up was a Manhattan. Before finishing it, he would fish out the two stemmed maraschino cherries my mother would put in the glass and give them to my sister Elaine and me. He later switched to California Zinfandel, which he bought in a jug and kept in the bottom cabinet of the china closet. Rest assured that, to a fourteen-year old, a water-diluted sip of Zinfandel was no match for the sweetness of a whiskey-soaked maraschino cherry.

I hadn't become a wine drinker until my early twenties. It was at a time when Mateus, Lancers, Blue Nun, and Chianti bottles in straw baskets were all the rage. Our neighbors in Anaheim had touted California wine and had encouraged us to visit the Napa Valley. It was on one of those early trips that I developed a feeling for the valley that resonated deep inside. Over the years, I found that being there provided a serenity and sense of harmony I didn't find elsewhere. Buying land there would make me part of the place.

I couldn't identify the origins of my sentiment for the valley. Both my parents were born and grew up in New York City, so my attraction to farming didn't come from them. But is it really necessary to fully understand why you like something when you do? I didn't think so. It was enough to know I liked it. After much soul searching, we decided that a vineyard had to be part of any property we bought because it would make us part of the valley's major business. Without it, we would be only tourists.

Patricia and I were having lunch one day at the Greystone Restaurant in St. Helena when a friend sent over a bottle of Schramsberg's sparkling wine named Querencia. A small booklet attached to the neck of the bottle explained that Querencia had been named by Jim Hickey for the Spanish word that conveys the deep and abiding affection each person has for the place he or she calls home: "It refers to the sense of being nourished by a place to which you belong." It was reassuring to know that other people had feelings similar to mine, because these words came as close as any I could use to convey the valley's appeal to me. Querencia was energizing my search for the right piece of real estate. Apparently the valley was a place that had been waiting for me to find it.

While the Napa Valley is my Valhalla, many others don't see it that way. It certainly does not offer the geographic scale and grandeur of the Grand Tetons, the cultural and artistic diversity of Santa Fe, the surging surf of Cape Cod, or the cosmopolitan appeal and excitement of New York City. While I enjoy those and other places, they don't provide the anchor for my psyche that Napa does. The good news is that we all don't treasure the same place.

As much as I wanted to buy real estate in the valley, there was a side of me that balked at the thought. I am part of a generation that still carries the taste of the Depression in its mouth. Not that I experienced it, but I grew up in a family that did, and I was influenced by what I heard. My parents were both born in 1912 and got married in 1934, two years after failures had closed one in every five banks in the country. While we never discussed it, I imagine the Depression caused them to trim their youthful expectations and perhaps even abandon some teenage dreams. Certainly the limited opportunities and deprivation of their times must have shaped their outlook about the future.

Their cautiousness was not lost on me. Despite the economic prosperity experienced by my generation, it was difficult to distance myself from those roots. Part of me felt it was frivolous for someone who lived in the Washington suburbs and had never sat on a tractor to want to own agricultural land, particularly a vineyard. Another part of me knew I didn't want to be drawing my last breath and wishing I had bought a vineyard just before I checked out for The Big Wine Tasting in the Sky.

Chapter 9:
The Napa Valley Gets Settled

Being on the verge, I hoped, of buying land in the Napa Valley, I found myself increasingly interested in its history. The creation of the Ag Preserve was part of the story, but what preceded it? What forces and what individuals forged the place I found so appealing? It's not possible to fully appreciate a place until you understand its origins.

While the valley's identity is tied to wine, it is a relative newcomer to the ranks of viticulture superpowers, having only achieved that status in the past few decades. By contrast, its counterparts in Europe like Burgundy, Bordeaux, Tuscany, and the Rhine have rich and exciting wine histories that date back centuries. Wine-writer Leon Adams suggested years ago that the romance and lore of those regions contributed to the taste and cache of the wines they produce. Napa, of course, has crafted its own history. It is one punctuated by economic cycles, natural disasters, and the nearly fatal man-made crisis of Prohibition. Its story reflects the persistence of visionaries whose grit and resourcefulness created the world-class California wine industry we know today.

The original inhabitants of the valley were people who relied on the plentiful water, fertile soil, and abundant game for their existence. Then, in June 1823, a Mexican expedition headed by Padre Jose Altimira from the Mission Dolores in Yerba Buena (now San Francisco) entered the valley in search of a new mission site. He met the people from the Napato settlement and christened the area Napa Valley, using what he understood their name to be. Eventually, however, he chose Sonoma as the place for a new mission and began construction in August of that year. When the Mexican government secularized the missions in 1833 and distributed their lands, the Sonoma Mission controlled an area of seven hundred square miles that included the Napa Valley.

As part of the secularization, a young military officer named Mariano Vallejo was selected by Governor Jose Figueroa to establish a presidio and make land grants to accelerate the population of the region. A presidio was a fortress used to protect the mission, its inhabitants, and its holdings. Secularization had increased tensions between the Mexicans and the native population, known today as Wappos, and skirmishes between them had become more frequent and bloody.

The native population was first diminished in number by warfare and later succumbed to epidemics of smallpox and cholera, diseases to which they had no natural immunity.

Although the Wappo no longer exist as a people, their legacies remain. Among them are paths they created that we call Silverado Trail and Spring Mountain Road. Many of the Wappo names have been anglicized and are still in use today, including Napato (Napa), Mallocamas (the Mayacamas Mountains that form the valley's western border), Suscol (Soscol is the major north-south road through the city of Napa), and Kaimus (both a land grant and a winery are named Caymus).

The first white settler in the valley was George Yount. A frontiersman born in North Carolina, he performed carpentry work for General Vallejo at the presidio, and Vallejo offered to pay him with the most abundant commodity at his disposal— land. The only catch was that Yount had to become a Mexican citizen and a Catholic. Apparently George was a pragmatist who had tired of his life as a trapper and hunter, and he accepted the offer. He was baptized at the Sonoma Mission and in 1836 received a land grant of about 11,800 acres in the central part of Napa Valley that now encompasses Yountville, Oakville, and Rutherford. He named it Rancho Caymus after the Indian settlement and moved onto it in 1838. Yount built a cabin on what today is Yount Mill Road within the boundaries of the community that bears his name, and shortly afterward he planted grapes. As history would prove, no place in the state would match Napa Valley's suitability for them.

Three years later, in 1841, an English physician named Edward Bale married one of General Vallejo's nieces, converted to Catholicism, and received a land grant bordering Yount's to the north. Bale built a home south of St. Helena but is best remembered for his mill. Located a mile north of St. Helena, Bale Mill is a California State Landmark and the oldest structure in Napa County.

As settlement began, word of the Napa Valley's desirable climate and land gradually began to spread. Farming represented the backbone of the economy from the outset. Livestock was raised on virtually all farms, but wheat was the principal income-producing product. There was a ready market for it and it could be transported anywhere in the world.

It was James Warren who began to promote viticulture in California in the mid-nineteenth century. As editor of the *California Farmer* and founder of the California State Agricultural Society, his views were influential. John Patchett, an Englishman, planted the first vineyard of any consequence in what is now the city of Napa, and in 1859 he also built the first winery. Despite the trailblazing efforts of Patchett, German immigrant Charles Krug is recognized as the first major viticulture figure in the valley.

In the late 1850s, Krug was working in Sonoma with Agoston Haraszthy, the most ardent and colorful promoter of California wine of his time, when he suddenly moved to the Napa Valley. The reason? Krug had fallen in love with Caroline Bale, the daughter of the late Dr. Edward Bale. As a result of her father's death, Caroline was due to inherit a sizable portion of Rancho Carne Humana upon marriage. When Charles and Caroline recited their nuptials on December 26, 1860, Napa Valley winemaking took a big step forward.

Krug planted grapes on land just north of St. Helena that had been part of Caroline's 540-acre dowry. It is the site of the Charles Krug Winery today. A plaque placed on the building by the California State Park Commission states that the winery was founded in 1861 and identifies it as the oldest operating winery in the Napa Valley. In 1862, another German immigrant, Jacob Schram, purchased property on Diamond Mountain just south of Calistoga and shortly thereafter planted the area's first hillside vineyard.

By the mid-1860s, the valley was home to two-dozen men who patiently planted, cultivated, pruned, and tended grapevines on a commercial basis and awaited their harvest in the fall. Krug and Dr. George Crane of St. Helena (after whom the park in that community is named) were among the most aggressive and ambitious of them and helped establish St. Helena as the region's wine epicenter.

Crane moved to St. Helena from San Jose and is credited with planting the area's first vinifera grapes, the species that produce the world's finest wines. Criolla grapes, usually referred to as Mission grapes, had dominated the area's vineyards until that time but produced wine of unpredictable quality.

Viticulturist Hamilton Crabb bought a 240-acre parcel in Oakville in 1868 and obtained cuttings from Bordeaux and the Loire Valley that flourished under his care. When his vineyard acquired a reputation for being the finest in California, he changed its name from Hermosa to To Kalon.

German-born Jacob Beringer arrived in Napa Valley in 1870 and was hired as Krug's winemaker. Six years later, he and his brother Frederick purchased land north of St. Helena for a winery. In 1883, they began construction of the elegant Rhine House that remains one of the area's most majestic landmarks.

While the Napa Valley was suffering through a depression that swept the entire country in the mid-1870s, the French wine industry was facing a disaster. An aphid-like insect named phylloxera was attacking the roots of the vinifera plants, stunting their growth and slowly killing them. French vineyards had never seen anything like it and were totally unprepared for the devastation that occurred. California took note, but was preoccupied with meeting the demand for their product caused by the French vineyard crisis.

The long-term prospects of the Napa Valley wine industry were given a boost in 1880 when wealthy Finnish sea captain and Alaskan fur dealer Gustave Niebaum purchased a large parcel of land in Rutherford. He planted vines immediately and began construction of his Inglenook Winery that was finished in 1887. With winemaking his avocation, he became the first of a long list of individuals drawn to the valley to invest in a winery using substantial wealth earned elsewhere. The magnificent structure he built houses the Rubicon Estate winery owned by Francis Ford Coppola, whose fortune is Hollywood-made and who also has European roots.

Steady improvement in quality and favorable publicity produced explosive industry growth in the 1880s. At the beginning of the decade, the valley had 49 wineries, but by 1886 there were 175. During this same period, wine production increased almost 60 percent.

Brothers James and George Goodman, sons of a Napa banking family, built the Eschol Winery in 1886 on a portion of the land owned by J.W. Osborne, one of the valley's wine

pioneers. The Goodmans were not Jewish, but on the suggestion of a friend they decided to name their winery Eschol, a
Hebrew word that means "valley of the grape." About the same
time, John Benson, who had struck it rich during the California gold rush, was building the Far Niente Winery a few miles
north of Eschol in Oakville.

The Eschol Winery eventually fell into disrepair after a
series of financial setbacks, and it was purchased in 1968 by
the Trefethen family. They restored its buildings and vineyards
and returned it to its position as one of the valley's premier wineries. Similarly, Gil Nickel purchased the rundown Far Niente
Winery in 1977 and took the painstaking and expensive steps
to rebuild its structure, caves, and vineyards. Both wineries
stand as examples of one generation of dreamers resurrecting
the creations of dreamers from an earlier day.

Another successful miner who tried his hand at winemaking was William Bourn. He and his partner selected a site on
a hillside just north of Krug, built a winery from native stone
cut at a nearby quarry, and named it Greystone. In later years,
it was operated as the Christian Brothers Winery and then
became the West Coast location of the Culinary Institute of
America that houses a contemporary restaurant offering fine
California cuisine.

Throughout this period of growth and expansion, however,
disaster was lurking in the soil. The insect that had bludgeoned
the French vineyards, phylloxera, was slowly spreading in
Napa. It was attacking the roots of the vine, injecting toxins
that sapped their strength and disrupted their uptake of water
and nutrients. It probably had been in the valley for twenty
years. A rootstock named V. californica had been widely
planted because it was considered to be resistant to phylloxera.
Unfortunately for California viticulturists, it was not and the
wine industry boom of the 1880s became a bust with three-
quarters of the vineyards being wiped out by 1898.

The curtain fell on the nineteenth century with the Napa Valley wine industry in a shambles. Thousands of acres of vineyard had been virtually abandoned to the ravages of a microscopic root louse. The future looked anything but promising. Fortunately, a leader emerged at the dawn of the twentieth century to change the momentum, restore confidence, and lead a viticulture revival. He was Frenchman Georges de Latour, who had immigrated to the United States in 1883 and purchased four acres in Rutherford in 1900. At that time, there were less than 5,000 acres of wine grapes in the valley, as opposed to over 18,000 in 1881. De Latour imported millions of benchgrafted Rupestris St. George rootstock from France in 1901 and found a ready market for them. Two years later, he purchased a parcel of 128 acres adjacent to the land he already owned and eventually incorporated his business as Beaulieu Vineyard, French for "beautiful place."

The replanting brought resurgence to the industry. By 1904, there were 7,000 acres of vines in the valley and the number rose steadily to 16,000 by 1910. Italian names like Domeniconi, Brovelli, Bartolucci, and Navone were becoming as commonplace as German names like Krug, Schram, and Beringer had been fifty years earlier. During this period, other forms of agriculture also prospered, and plum trees for the production of prunes actually accounted for more acreage than grapes and had a slightly higher crop value.

World War I took a toll on the French wine industry and benefited California wine sales, but the industry's momentum was jarred when Prohibition became law in 1920. The state-by-state approval process of that constitutional amendment provided ample warning of its approach; nonetheless, its finality literally shook the valley to its roots. In anticipation, acres planted in grapes gradually declined and those planted in fruit and nut trees increased. Softening the harsh effect of the law was the provision that recognized winemaking as one of the oldest of home arts and permitted families to make up to two hundred gallons of wine each year.

Despite Prohibition, Georges de Latour's business flourished, and it is reported that he became the wealthiest man in the county during this time. His Jesuit education in France had helped him befriend high-ranking clerics in the Catholic Church, and his relationships with the archbishops of San Francisco, Los Angeles, and New York helped him become selected as the major supplier of sacramental wine for many dioceses across the country. Beringer and Larkmead, a winery built in the 1880s, also produced altar wines, but Beaulieu dominated the market. The success of de Latour's winery provided work for those lucky enough to get it and made life much easier for him than it was for other winery and vineyard owners.

Just on the horizon, however, hovered economic clouds that would deepen the valley's economic woes.

Chapter 10:
Creating a Wine Tradition

The Napa Valley's economy was already reeling from the effects of Prohibition when it plunged into a deeper spiral because of the Great Depression. A report to the county's Board of Supervisors showed that 1930 income from fruit sales had dropped almost 75 percent from 1922 and amounted to only $1 million, with grapes representing one-third of that figure.

When Prohibition was repealed on December 5, 1933, the wine industry faced a challenge as daunting as any ever faced in this country. To begin with, Prohibition had driven the drinking of alcoholic beverages underground, where bootleggers controlled production and gangsters handled distribution. Americans had learned to live without wine or had become accustomed to homemade or bootleg wine that usually was coarse and harsh. Wine was a nonessential product to all but a small percentage of families whose European heritage made it an essential part of their meals. Promoting sales of such a product in the middle of the Depression was no small task.

Just as important was the fact that the workforce had become depleted, as experienced people were forced to leave the industry and the region in search of employment. They took with them essential knowledge about vineyard management and production methods. Replenishing the industry's talent pool was a process that could only be accomplished over several years.

Prohibition also stopped dead in its tracks the various forms of experimentation and innovation that are essential to progress and advancement in every industry. And, of course, the poor physical condition of winemaking facilities at Repeal meant that an extensive amount of new investment would be required to restore production capacity. Nothing less than a complete rebuilding of infrastructure combined with re-educating consumers about wine was required for the industry to get off its back. It had to virtually reinvent itself.

When Repeal was on the horizon, the Christian Brothers located a winery in the hills northwest of Napa in 1932. Louis Martini, a major vineyard owner in the San Joaquin Valley, bought ten acres south of St. Helena in 1933 and announced his plans to build a new winery and make the Napa Valley his home. The modern plant he opened the following year boasted a storage capacity of 1,250,000 gallons. The Bisceglia brothers began restoring the Greystone cellars they had purchased almost a decade earlier. Mrs. Niebaum reactivated Inglenook under the direction of Carl Bundschu, and Louis Stella attempted to put the Krug Winery back on its feet. The valley was re-energized with hope and bright prospects. Even nature cooperated with good weather that made the 1933 harvest a banner one.

Often an innovator, Georges de Latour began bottling and sealing his wine at the winery to ensure quality. This occurred at a time when 80 percent of wine was still sold in bulk to companies that blended, stored, and bottled it under their own label rather than the winery's. The assurance of quality offered by bottling further boosted Beaulieu sales by enabling it to sign a contract with Park and Tilford to market wines on the East Coast, where it was usually sold in bottles of a half-gallon or more.

The California Wine Institute was created in 1935 and Leon Adams, a San Francisco journalist, served on the Publicity and Promotion Committee, which took on the task of winning back the consumer. Albert Winkler became the chairman of the new Department of Viticulture at the University of California's Davis campus near Sacramento in 1936. Previously, all the work related to wine had been done at the Berkeley campus. Winkler hired Maynard Amerine into his new department as a professor of viticulture. Amerine's work over the next fifty years earned him recognition as America's most esteemed enologist. The faculty at UC Davis worked diligently to convince the industry that science and technology played important roles in farming and fermentation and in the process made an education from UC Davis valued and prized throughout the world of wine.

More than forty wineries were operating in the valley by 1937. Among the many Italians who had settled there was Cesare Mondavi, a grape grower from Lodi who became a major partner in the Sunny St. Helena Winery. Mondavi's sons, Robert and Peter, began learning the business from the ground up and preparing for their prominent future roles in the industry.

In 1938, de Latour traveled to the Pasteur Institute in Paris, seeking assistance from a friend on the faculty. His wines frequently experienced a chemical breakdown, and he needed help in diagnosing and correcting the problem. The professor recommended one of his students, Andre Tchelistcheff, a Russian émigré who had distinguished himself in experimental viticulture and the science of fermentation. De Latour convinced Tchelistcheff that Beaulieu would be a wonderful place to practice his skills, and their handshake agreement to have Andre join Beaulieu had a profound effect on the course of winemaking in America.

Tchelistcheff was born in Moscow and grew up on his father's country estate. The family was forced to flee Russia during the Revolution, and Andre received training in viticulture and winemaking in Hungary. After a failed business venture in France, he resumed his winemaking studies at the Institute of National Agronomy and took courses at the Pasteur Institute that brought him to the attention of de Latour.

It was Louis Pasteur who had identified the principles of fermentation in 1860 and had become the world's leading authority in the emerging field of microbiology. He brought the rigor and discipline of science to the ancient art of winemaking, changing it for all time. The Russian with the unpronounceable name knew that hygiene was crucial because unwanted bacteria and mold can devastate wine. He imposed completely new standards for cleanliness and technical precision upon his arrival at Beaulieu and introduced winemaking innovations like cold fermentation that improved wine quality. Tchelistcheff solved the problem that had taken de Latour to Paris, and Beaulieu's reputation soared. Sound technology and creative marketing were the engines that drove Beaulieu's reputation for quality for the next forty years.

At the same time, East Coast wine importer and writer Frank Schoonmaker was developing an interest in California wines. He and a handful of others promoted the idea that the state's wines should be identified by grape variety, such as Sauvignon Blanc, Merlot, and Cabernet Sauvignon, and specific geography, such as Oakville and Mount Veeder, rather than by European-place appellations like Burgundy or Chablis. His concept was resisted by most wineries, since it required everyone in the chain of consumption— wholesaler, retailer, and consumer—to become educated about varieties and locales. Many producers of fine wines, however, recognized that varietal and geographic labeling could become the basis for differentiating their products and supported the adoption of it. In retrospect, this was one of the most influential concepts used to transform the Napa Valley wine industry from one that produced a bulk commodity to one that produced branded products recognized and respected around the world for their quality.

With the death of Georges de Latour in 1940, Tchelistcheff, Martini, Bundschu, brother Timothy of Christian Brothers, and Felix Salmina of Larkmead became the guiding lights of the industry. However, it was John Daniel, the grandnephew of Mrs. Niebaum who inherited Inglenook upon her death in 1936, that set the valley's standard for wine excellence during the 1940s and 1950s. Through his relationship with de Latour, Daniel got assistance from Tchelistcheff and earned his respect for pursuing quality with a vigor most others did not demonstrate.

During World War II, the unavailability of tank cars and price control regulations discouraged the production of bulk wines and further pushed Napa Valley winemakers into becoming producers and marketers of bottled and branded wine products. It was during the industry's boom near the end of the war that Cesare Mondavi bought the rundown Charles Krug Winery and 160 acres of vineyard for $75,000. He had been urged to do so by his son Robert, who recognized the opportunity to buy a winery of historic significance and use it to create a family brand. In this same period, Martin Stelling Jr. bought a portion of the old To Kalon Vineyard together with hundreds of adjacent acres and planted high quality varietals. It is almost certain that his decision was influenced by Professors Winkler and Amerine, who had been advising vineyard owners that Cabernet Sauvignon and other quality varietals would produce better wines than the Alicante and Palmino grapes that still dominated the valley. Stelling was killed in an automobile accident in 1950, but his plantings were a legacy to others in the quest for higher-quality wine. Others followed Stelling's lead, and the willingness of growers to plant relatively low-yielding, high-quality varieties was a turning point in Napa's viticultural history.

Sales of wine grew slowly and steadily during the 1950s and 1960s. The push for quality was being driven by changes in consumer tastes and the dedication of forward thinkers like Tchelistcheff, the Mondavis, Lee Stewart of Souverain (now Burgess Winery), the Taylors of Mayacamas Vineyards, and the McCreas of Stony Hill Winery.

The Napa Valley wine industry was adjusting to the changes in consumer tastes that prosperity, education, and world travel were making in the buying and drinking habits of Americans. Quality was the driving force, and marketing was increasingly crucial to winery success.

Cesare Mondavi died in 1959, leaving his wife Rosa in charge of Krug. Without their father to manage them, Robert and Peter were increasingly at odds over important winery decisions. Finally, a fight between them in 1965 led to Robert's ouster from the winery by his mother. At age 52, he suddenly found himself estranged from his family and without a connection to the industry that had been his life's work for more than thirty years. Following a brief hiatus, Robert found financial backers, bought part of the To Kalon Vineyard in Oakville, and built a winery that featured early California mission architecture. It was the first one designed to attract visitors. In the process of this and future changes, Robert Mondavi, who passed away in May 2008, demonstrated the vision and qualities that made him the icon of American winemaking.

In *Harvest of Joy*, Robert recounted a trip to Europe in 1962, which gave him insight into the know-how and traditions of the great European chateaus. He left the continent elated. The soil, climate, and rootstocks in the Napa Valley were as good as, and perhaps better than, those he had seen in Europe. He was certain that if he and his colleagues had the passion, conviction, and willingness to invest in research, development, and new equipment, they could make wines equal to the most heralded wines of Europe. He returned to the Napa Valley— a place he described as a sleepy little farm community with no sophistication, little economic vitality, and no fine restaurants—energized and ready to go. His energy in pursuit of this vision, his trailblazing behavior, and his capacity to engage others in the quest to raise quality standards were relentless.

By his own admission, Robert Mondavi was not quiet, re-tiring, or self-effacing, yet he was unquestionably a winemaker, farmer, sage, scientist, marketer, Pied Piper, and P.T. Barnum rolled into one. More than any single person, he was responsible for shaping the wine industry found in California today. Some observe that he was prone to brag a bit too much about his accomplishments. When accused of the same thing decades earlier, Dizzy Dean, the star pitcher of the St. Louis Cardinals, observed, "If you can do it, it ain't braggin'." And Mondavi had done it. He led the industry's transformation from a bulk wine producer to one characterized by highly differentiated brands with reputations for excellence and price tags to match. He promoted innovation in farming, winemaking, and market-ing and never wavered from a commitment to excellence. He challenged the industry to raise its sights and think and act differently. He set the direction and led the way.

In September 2001, Mondavi was honored at a dinner held by the Oakville Winegrowers. Dennis Groth, Clarke Swanson, and Jean Phillips were among the notables who paid homage to him. The sentiments expressed were best summarized by Augustin Huneeus, a long-time friend and acquaintance, who observed, "We value your generosity, Bob. But most impor-tantly, you have been an inspiration and have taught us that sharing our knowledge is what takes all of us forward."

There is no debate about the event that catapulted the Napa Valley into the ranks of the world's premier viticulture centers. It occurred in Paris on May 24, 1976. An Englishman named Stephen Spurrier, who owned both a wine shop and Academie du Vin in Paris, decided to sponsor a blind tasting of California and French wines. His visit to California the previ-ous year had convinced him that something exceptional was happening to winemaking there. He concocted the idea of a tasting as a way to promote his store and school.

Spurrier arranged to have nine experienced, widely known, and highly respected French wine judges gather at 3 p.m. in the central courtyard of the fashionable Inter-Continental Hotel to taste twenty wines. A small group of onlookers enjoying the warm, pleasant afternoon gathered to watch. It included only one journalist, George Taber of *Time* magazine, since no one in the French press saw it as an event worth covering.

The bottles were all covered to disguise them. Ten of the wines were whites made from Chardonnay and ten were reds made primarily from Cabernet Sauvignon. In each category, six were American made and four were French made. To en-sure a valid standard of comparison, Spurrier had chosen fine French wines. The French Chardonnays, or white Burgundies, were represented by one Grand Cru and three Premier Cru. The French reds were represented by two First Growths and two Second Growths from the 1855 classification of French wines. All were predominantly Cabernet Sauvignon Bordeaux blends. Both the white and red California wines were from wineries started or reopened in the 1960s and early 1970s.

The judges proceeded to observe, sniff, swirl, taste, and spit as the spectators watched. When their ratings were completed and Spurrier unwrapped the bottles, the highest-rated wine in each category was American—a 1973 Stags Leap Wine Cellars Cabernet Sauvignon and a 1973 Chateau Montelena Chardonnay. Every French judge rated a California Chardon-nay first. The red wine competition was closer with Stags Leap just edging out Chateau Mouton Rothschild. Taber's article about the tasting appeared in the June 7, 1976, issue of *Time* and set the wine world on its ear. It closed with a quote from Jim Barrett, Chateau Monthelena's general manager and part owner: "Not bad for kids from the sticks."

Stags Leap Wine Cellars was the handiwork of Warren Winiarski who had come to the valley in the 1960s from Chi-cago with an interest in winemaking. He worked with Lee

Stewart at Souverain and then Robert Mondavi before buying forty-five acres east of the Silverado Trail. The winery he opened in 1972 was named for a nearby outcropping of stone referred to as Stags Leap.

Chateau Montelena, originally built in 1886 by San Francisco businessman Alfred Tubbs, had been sold to a partnership headed by Jim Barrett in 1969. They refurbished it and hired as winemaker Miljenko (Mike) Grgich, a Croatian who had come to the United States by way of Canada. He had previously worked at Souverain, Christian Brothers, Beaulieu, and Mondavi. The prize-winning Chardonnay was only their second vintage!

In addition to being outstanding winemakers in their own right, both Winiarski and Grgich were the beneficiaries of know-how and experience that had accumulated over decades of winemaking in the valley—matching grape varieties to particular soils and climates, improving grape quality by proper exposure of leaves and grape clusters to sunlight, using manufactured yeasts, controlling temperature during fermentation, using new methods to press grapes, using both oak and stainless-steel barrels, and on and on.

Neither Winiarski nor Grgich attended the tasting, and neither was aware that their wines had been entered in the contest. Discussing the occasion during an interview a few years later, Winiarski observed, "I think if you look at the results and try to understand what was really important, it was not that our wines were chosen above some French wines. That was gratifying, pleasant, competitively significant, and all those things, but the fact that testers could not systematically separate our wines from their wines meant that we had achieved a certain classic character."[1]

---

[1] Warren Winiarski, "Creating Classic Wines in the Napa Valley," an oral history conducted in 1991, 1993 by Ruth Teiser, Regional Oral History Office, The Bancroft Library, University of California, Berkeley, 1994, pp. 67-68.

From a business point of view, the benefit was not esoteric but real. When he was interviewed for the April 26, 2001, issue of the *Yountville Sun*, Winiarski said, "After the Paris tasting, I no longer had to call dealers and distributorships. They called me. That was a total change. You have no idea how difficult it was to sell California wines before the tasting."

The Napa Valley was granted status as an appellation by the Bureau of Alcohol, Tobacco, and Firearms (BATF) in 1983. Subsequently, regions within the valley (such as Oakville and Rutherford) have been designated as appellations based on factors like soil, patterns of climate, elevation, rainfall, and reputation as a wine region. According to BATF regulations, a label containing an appellation designation means at least 85 percent of the grapes that produced that wine must have been grown and harvested in the appellation.

The 1980s, however, also saw the reappearance of phylloxera. It attacked the AXR-1 rootstock, which the University of California at Davis had been promoting for decades because of its ability to produce fine grapes and resist disease and phylloxera. But AXR-1 had not failed as a phylloxera-resistant rootstock as much as phylloxera had evolved into a new, nastier strain labeled Strain B. The bright side of this disaster was that it enabled vineyards to replant with varieties that were well suited for their soil and climate and to use the most current knowledge about rootstock, vine spacing, and trellising.

A trend that first appeared in the 1980s and then accelerated in the 1990s has been the valley's emergence as a culinary center. Robert Mondavi had grown up in a family where wine was not a stand-alone beverage but one that enhanced meals. Thus, at Robert's winery, he promoted the idea that wine should accompany food. This renaissance in winemaking paralleled the emergence of California cuisine emphasizing fresh and healthful ingredients and dishes that are wine friendly and wine inspired. Alice Waters's Chez Panisse Restaurant in Berkeley became the benchmark for this style of cooking, and it inspired others. The concept caught fire just as the Napa Valley was becoming an increasingly popular tourist mecca. From the classic four-star menu at the French Laundry to chic and tasty fare offered at dozens of other fine restaurants, the valley has become known for its excellent food as well as its world-class wines.

Indicative of this transformation was the 1995 opening of the Culinary Institute of America's West Coast campus at the former Christian Brothers Winery, two miles north of St. Helena. This institute attracts thousands of food and beverage industry professionals to dozens of programs each year and houses a restaurant that features an airy, open setting and a fine menu.

As one of the premier winegrowing regions in the world, Napa Valley goes head-to-head with other famous regions for the attention of consumers. It received an unexpected boost in that regard from wine critic Robert Parker who lavished praise on it in the October 31, 2013, edition of his publication *The Wine Advocate*. Commenting on the interest shown in the Napa Valley by some of the notable families in French wine-growing, he wrote, "They recognize the consistency and brilliance of vintages from year to year, as Napa harvests are far more uniform and consistent in quality than Bordeaux. Most importantly, world-class quality from Napa Valley is a fact, not a myth. All the fallacies about Napa wines being too rich, short-lived, and over-the-top are, in fact, absurd drivel often created by Euro-centric wine drinkers that have been proven false time and time again. If you are not following what is going on in Napa Valley, you are missing some of the world's most compelling and provocative wines. End of story."

Chapter 11:
Sharpening Our Focus

The Napa Valley location of our dreams was proving to be elusive, but persistence would be our ally. On the basis of eight trips, I had become an experienced traveler between Washington Dulles and San Francisco International Airports. United Airlines dominates that route and I was getting well acquainted with its service.

At cruising altitude, no single feature on the ground dominates your view, and your eyes naturally roam over the landscape. You develop an appreciation for the vastness of our country and realize how open and sparsely populated most of it is. Life below is a miniaturized version of the real thing and moves at a leisurely pace. From an airplane window, you realize that lines in nature curve and are irregular and that forms are rounded. Straight lines are usually associated with man-made creations like roads, boundaries, and buildings.

Colors are softened by the filter of five miles of atmosphere. The greens of forests range from almost black to nearly white, and earth tones range from tan to gray and from rose to maroon. With the exception of clouds and snow, white objects are almost always crafted by human hands. But the most astonishing thing about flying to me still is the fact that someone figured out how to get hundreds of people and tons of equipment off the ground and up into the sky and move us at a speed of six hundred miles per hour through the air. Could Lewis and Clark even comprehend such travel?

Patricia and I marked the sixteenth month of our search by visiting friends in Bodega Bay on the Pacific Coast about an hour and a half north of San Francisco. The drive from the airport took us through a picturesque part of rural California we had not visited before and was a welcome variation to our schedule. The ever-expanding real estate file that included the current issue of *Distinctive Properties of Napa Valley* was stashed in my luggage. Glancing through it the first night at Bodega Bay, one of the ads caught my eye:

> *Currently licensed as a 2 bedroom Bed and Breakfast. Can be converted to 4 bedroom, 4 bath home. Property is about 6 acres with personal winery and 3 acre vineyard …*

The location was not identified, but we thought we had seen it before.

"That looks like the right size for us," Patricia observed, looking at the photo that accompanied the ad. "The house isn't overpowering like Byrd Hill, and six acres seem manageable."

Strong winds blowing off the Pacific made Bodega Bay less pleasant than expected and gave us a reason to return to the valley sooner than planned to see this new possibility. The owner had already moved out, and Steve had no trouble getting access to the property. It was a two-story house with a stained-wood exterior on Lodi Lane. The landscaping was overgrown, and the interior showed signs of age. A do-it-yourself addition to the original structure built by a prior owner created a choppy floor plan, yet it felt comfortable. A large building across the gravel driveway from the house could be used for storage or, as the ad said, a personal winery. The vineyard behind the house was unimpressive. Despite the drawbacks, Patricia liked the feel of the place.

"The scale is much better for us than Byrd Hill or any of the other parcels we've seen," she volunteered. "It's the kind of location I was hoping to find." By midafternoon, we were in

Steve's office to discuss an offer.

"I never figured you for that place," Steve observed.

"Why?" I asked.

"I don't know. Maybe because it's so simple. Or maybe because it's not off by itself like so many places you've liked."

Our offer provided for a thirty-day period to complete all inspections to our satisfaction. The counteroffer was reasonable, and within twenty-four hours we came to terms with the owner subject to the satisfactory completions of inspections. I dropped in on Tom Paine and Teresa Norton to tell them the good news.

"Why would you want that place?" was Teresa's straightforward question. "It gets flooded. This past winter it was inaccessible because of high water."

"Do you know anything about the soils?" Tom asked.

"Not yet. My realtor, Steve Ericson, has recommended someone in the vineyard management business to look at them, but I'd appreciate any recommendations you have."

"I'd suggest David Abreu, who manages our vineyard. He knows this part of the valley as well as anyone and may have an opinion."

Tom gave me David's number, and as I finished writing it down I looked up to see both Teresa and Tom looking at me quizzically. Rather than sharing my excitement, they were restrained, even critical. Their body language and their words were unmistakable. They didn't think this was a good buy.

I called David the next day and explained who I was. "I was expecting your call," he said pleasantly. "Tom spoke to me this morning."

"Do you know the parcel I'm considering?" I asked.

"Sure do," he answered.

"What are your thoughts?"

"How candid do you want me to be?" he inquired.

"As candid as if you were talking to your best friend."

He paused. "Anyone who buys that place would do better grazing horses and cows on it than growing grapes. There's too much clay. You can't get anything good to grow there."

David obviously was not one for mincing words. I asked a few more questions, but he had given me a clear picture. It raised a question we had never discussed. Did we want any kind of vineyard or did we want one that would produce quality grapes? That required some thought.

My tour of the vineyard with the vineyard manager Steve recommended was informative. Like Steve, he had grown up in the valley and had a successful vineyard-management business.

"The land here is okay," he volunteered. "One of the three planted acres has good soil. The other two acres have too much clay. The vines are not healthy and I'm not sure why. You'll probably have to replant."

The next morning I went to Albion Survey to request a survey of the elevation of the house relative to the flood plain. When I finished talking to Rob Bell, the owner, another customer standing nearby who overheard our conversation remarked, "I hope you stole that parcel."

"Why's that?" I asked.

"It floods and the soil's no good. If you didn't get it at a real good price, you don't want it." He smiled and walked past me out the door. I looked at Rob, who shrugged his shoulders.

"Still want me to check the elevation?" he asked.

"Yes," I answered. "I still want to know."

The karma was not good. Starting with the reaction of Tom and Teresa and continuing with David Abreu's comments, the observations of an experienced vineyard manager, and the unsolicited opinion of an anonymous Albion Survey customer, the messages I was getting were all negative. Patricia had expressed stronger support for this parcel than any of the others, yet it was apparent that the disadvantages were swamping the advantages. The location was a plus. The existence of both a house and a vineyard meant we wouldn't have to start from scratch. The disadvantages were more extensive. It was located in a flood plain, it would need substantial renovations, and the vineyard contained only marginal soil and would have to be replanted. In addition, no one's first reaction had been positive, including Steve, who stood to earn a commission from the sale. The final blow came the next day with the disclosure that the elevation of the house was within a foot of the flood level, not leaving much margin for error.

Patricia shrugged her shoulders when I finished summarizing the pros and cons.

"If it isn't right, it isn't right," I said. "We can't force it."

And with those few words, we put a hold on our search. It was time to give it a rest.

Chapter 12:
An Unexpected Challenge

When you begin an adventure like this, you have no idea when it will end and you can't really be sure how it will end. Of course, you expect it will end with you owning a vineyard, but what if that doesn't happen? Suppose after searching for a year or two you decide for some reason it's really a bad idea. Maybe you discover that you romanticized it and, when the fog cleared, you realized it's too risky, too expensive, or more work than you ever expected. Maybe you'd be better off continuing to buy wine off the shelf of the local liquor store rather than trying to immerse yourself in the wine industry.

Because the ending is uncertain, you have to decide what to tell your friends about what you're doing. Do you even want to say anything in case the idea goes belly up? If you don't tell anyone, then you never have to explain why you changed your mind or why it didn't work out. What made me particularly wary about what to say grew out of an experience that occurred when our plans were still in the formative stages and we had not yet made a firm decision to proceed. I was on a driving range in Bethesda, Maryland, hitting golf balls one lovely afternoon when a friend came over and began to share with me some retirement plans he was making. He and his wife were considering a move to Kiawah Island, South Carolina, which had been their favorite vacation place for years. It was a big decision, he explained, and, since it was not final, he asked me to keep their little secret to myself.

I assured him I would and said I appreciated the fact that he shared that confidence with me. What an opportunity, I thought, to tell him about the plans we were considering. You know, one buddy confiding in another.

"Your plans sound great, Rick. I know how much you like Kiawah. Patricia and I are nearing the end of our corporate life and are considering some alternatives, too."

"Terrific. What do you have in mind?" he inquired.

"We're thinking of buying a vineyard in the Napa Valley. It's a place ..."

"You're what?" he asked incredulously.

"We're thinking of buying a vineyard in the Napa ..."

"That's the stupidest thing I ever heard of, Cliff. What do you know about vineyards?"

"Nothing. But I can learn and it's a special place to us like Kiawah ..."

"Cliff, for heaven's sake, think of what you're saying." By this time, his voice had gone from the slightly-above-a-whisper voice he used to tell me his little secret to a voice that could increasingly be heard by others nearby who were trying to concentrate on their turn, the position of their hands at take-away, and the fifty other variables that comprise a good golf swing. One guy in particular looked over with a disgusted, "You know this is a driving range and not a bar," look on his face. But Rick was undeterred. "You want to own a business you know nothing about that is 3,000 miles across the country and takes one full travel day to get there. That doesn't make sense."

"Those are some of the factors that have kept us from making a final decision but we're close," I said softly, trying to return our conversation back to a low decibel level.

"For heaven's sake, don't do it," he said in a raised voice. "You'll hate yourself for making such a mistake. I don't know how you could come up with such a harebrained idea."

By now, my head was spinning. My private tete-a-tete had turned into a public dressing down. Rick, for whatever reason, was at his judgmental best, criticizing our plans without ever inquiring how we had come up with them. And doing it in a rather loud voice. Finally, in what could only be described as disbelief, he hit his forehead with the palm of his hand and walked away.

So much for confiding in friends, I thought. Fortunately, no one had heard enough to know what happened. Or so I thought until I was in the Pro Shop thirty minutes later and another friend walked in and said, "Hey, Cliff, I hear you and Patricia bought a vineyard. Where is it?"

"Where did you hear that?" I asked.

"From George. He was hitting balls and overheard you talking with Rick."

I assured my friend he was misinformed and that if we ever bought a vineyard he'd hear about it from me and not through the rumor mill.

What I had not expected was for someone to so outspokenly condemn buying a vineyard by saying it was a bad – no, a stupid – idea. I expected people to make an innocuous comment or two or just shrug, or maybe even, for the sake of a friendship, say it was a neat idea. Rick's reaction was food for thought. He had nailed the serious drawbacks. What if he was right?

Once we started our search, the time I was spending in California was noticed by colleagues and I felt the need to explain to them what I was doing. They knew Patricia and I enjoyed wine and had lived in California, but I'm not sure anyone thought we were really serious. They'd ask for updates, but after hearing about the near-misses at Byrd Hill, Taplin Road, and Lodi Lane, I was getting the impression they thought the prospects of the Ehrlichs ever owning a vineyard were about as likely as my handicap dropping from eighteen to eight.

Chapter 13:
Still on the Prowl

My time, travel, and effort had produced nothing more than a friendship with Steve Ericson and volumes of knowledge about real estate in the Napa Valley. I was in the peculiar position of liking the Napa Valley but not the property there that was for sale. Were Patricia and I being too picky? Maybe, despite the influence of querencia, we were meant to visit and enjoy rather than own. The words of my friend who pointed out the economics of visiting rather than buying were rattling around in my brain.

When you get swept up in something as emotional as this, it's difficult to disengage. Your goal-oriented side wants to keep going because the goal has not yet been attained. But in the scheme of things, how important is the goal? I had already lived almost sixty years without owning land in the Napa Valley and had found life to be just fine. Nonetheless, I felt the need to search long enough to determine whether there was a parcel Patricia and I could enjoy. Which led to the question: How long is long enough? I didn't know. The fact that my enthusiasm was not diminished by the trips and disappointments told me I had not yet reached the saturation point. When it was time to throw in the towel, I would know it.

It was not necessary to make any earthshaking decisions. I could just change the frequency of my trips. Instead of every five to six weeks, I could visit every three to four months instead. We were not in a rush. The only pressure was coming from the market. With prices rising, the longer we waited the more expensive we could expect our purchase to be. The valley appeared to be a hot destination, but I wasn't sure how much of that view was magnified by my own anxiety.

An article that appeared in the July 18, 1997, issue of the *Wall Street Journal* provided a sanity check. The headline announced "Nouveau Vinters Toast Napa's Real Estate Boom" and read in part:

> *A booming economy is driving professionals to invest in second homes in America's most prestigious wine-growing region, while soaring grape prices and increased consumption are helping revive the wine industry. The combined effect: a real estate boom in the valley about an hour's drive north of San Francisco.*

> *The Napa Valley is so hot that houses with vineyards— from estates to fixer-uppers—don't last more than a few days on the market. Local real estate agents estimate that the dollar volume of their business is up almost 50% this year and say the biggest problem is a shortage of inventory.*

Buried in the middle of the article was a reference to a parcel that could only have been the Bartlett property on Crystal Springs Road that had been bought out from under us:

> *In the hills above St. Helena, a family is selling a large tract purchased in the 1950s ... The first parcel went ... to a San Francisco radiologist who is developing vineyards and building an 800-square-foot house out of remilled barn lumber and wine barrels.*

I faxed a copy to Steve. The country was in the middle of the dot-com boom, and reading the article gave me a mental image of hordes of well-heeled, blue-jean-clad Silicon Valley buyers racing up Highway 29 in their BMWs and storming into real estate offices, checkbooks in hand. By contrast, I was in my second year of plodding. If I put my search on hold for very long, I ran the risk of the hordes picking the carcass clean and leaving me the leftovers. That was a dismal prospect. Not on your life, I thought. Searching at a more leisurely pace would not be in my best interest. I was not going to get pushed aside by a generation of newly minted millionaires who weren't even around when Patricia and I ventured into the valley for the first time in 1967. I looked at my schedule, pulled out my airline guide, and called Steve to make arrangements for my next visit.

He was, as always, encouraging. "I found a couple more. We've looked at one of them already. It's at the end of Ehlers Lane. There are three parcels. You really should give one or two of them serious consideration. And I heard there might be another parcel coming on the market. It's owned by a man who has been talking about selling for the past three years and has decided now may be the time." He paused and then added, "There's always something."

Always something. That's what kept me and the Silicon Valley hordes coming back. Their advantage was having to travel a shorter distance.

Steve and I met at Gillwoods for breakfast on my next trip. It's one of the "in" places in St. Helena where locals gather for breakfast and lunch. Located across Main Street from Steve's office, it has a very small waiting area and tables for twos and fours squeezed close to one another. Watercolors and oils by local artists available for purchase decorate painted walls. A community table located on the other side of the windshield from the door provided seats for six or seven on a first-come, first-serve basis. Steve recognized some of the people sitting there and we joined them. His introduction of me prompted an assortment of nods and greetings. One of the diners was an attractive, perky woman who looked familiar. Steve introduced her as Robin Lail.

I really didn't know her, but I knew of her. She was one of John Daniel's two daughters who was reported to have inherited a portion of the Inglenook vineyard when he died. They had formed a successful joint venture named Dominus with Christian Moueix of the world-famous Chateau Petrus. Her husband was one of the valley's prominent architects.

Amidst the multiple conversations going on around the table, Steve explained to her that I first came to the Napa Valley in the late 1960s, had been coming back regularly, and had been looking to buy something. Somehow this message got garbled in the transmission, because it caused Robin to put down her fork and look over at me with a shocked expression.

"You've been trying to find property here since the 1960s?" she asked. Rolling her eyes, she added, "My God, Cliff, loosen up! Don't be so picky."

I laughed and corrected the facts, but her comment was food for thought. Loosen up. Be less picky. It sounded right, but if I followed it we would have owned something that was marginal. It was better, I thought, to be picky and keep grinding.

After our juice, pancakes, and coffee, Steve and I were back on the road. We looked at the property at the end of Ehlers Lane again, but I couldn't warm to it, and then we visited two more parcels in St. Helena before Steve mentioned a listing that had caught his eye that morning. It was a three-bedroom house in Oakville on ten and a half acres, including a seven-acre vineyard. It had come on the market in early June but was priced well outside my price range.

"Take a look," he said. "It's higher than you want to spend, but Oakville is a great area and you should get some idea of what's available south of St. Helena. I made an appointment to meet the listing agent there right before lunch."

Our trip down the Silverado Trail and across tree-lined Oakville Cross Road eventually brought us to a two-story house that the listing said contained 2,500 square feet. It was situated in an orchard of walnut trees with a vineyard behind it. Other vineyards occupied the land on either side. It probably had originally been painted a barn-like red, but years of sun and rain had changed the color to dark reddish brown. Weeds blanketed the ground under the trees and partially hid a pile of old construction beams, a rundown piece of farm equipment, and an assortment of wooden and metal items. The owners looked to be compulsive savers—the kind of people who believed everything they owned would have a practical use someday.

The realtor who listed the property met us in the gravel driveway. He and I walked up five steps to the front porch and went inside. The owner's teenage daughter was the only person at home. She greeted us and then quickly disappeared into her room on the second floor. The first floor had a kitchen, master suite, office, and family room with windows that faced north and overlooked the vineyard. Two bedrooms, a sitting room, and a bathroom were on the second floor loft. The wide, covered front porch wrapped around the west side of the house and opened into a large deck elevated about three feet above the ground. All things considered, this property was the best I had seen in almost two years of searching.

The steps had prevented Steve from making the tour, so when the other realtor left, he pumped me about the interior and I explained the layout.

"What did you think?" he said when I finished.

"It's a great spot, Steve, but the price blows by the upper limit we set. It was good to have the opportunity to see it."

"She'll never sell at that price," he remarked. "I wanted you to see it for when she lowers the price."

"In this market, she may sell it in a week."

"Maybe, but I don't think so."

Set almost in the middle of the valley, there were unobstructed views to the west, north, and east. Unlike the property on Taplin Road, there were no power lines to impair the scenery. The house was off by itself, but not remote. While it needed some fixing up, it was well designed and comfortable. A county-assessor map Steve had copied showed that the parcel to the west was a house and vineyard on sixteen acres, while Heublein owned a new vineyard planted to the east and north. A well-cared-for home and vineyard occupied the land to the south.

Then we were off to look at a twenty-acre Zinfandel vineyard behind Sattui Winery and a small parcel behind Corison Winery. I watched carefully as we drove around but saw no signs of the dot-com crowd swarming over the landscape like locusts. No real estate gems had been uncovered, but I could return home confident that a plodder like me still had a chance to find something nice in the Napa Valley.

Chapter 14:
The Weather Was No Drawback

In 1997, I retired from a twenty-five-year career at Marriott with a wing-ding of a party compliments of my boss, Bill Marriott, who is as skillful and inspirational a business leader as there is in America, and my assistant, Nancy Johnson, who arranged it. Included among my gifts was a magnum of 1992 Opus One in an engraved bottle and a two-night stay at the Mondavi Guest House. Patricia and I decided to use it in September, a month in which the Bay Area could be counted on for good weather.

The experience was every bit as good as we had hoped. The guest house was located in Stags Leap on a knoll set back almost a quarter of a mile from the Silverado Trail. A one-story building originally built in the 1930s, it had been stylishly renovated and decorated with a Laura Ashley touch. High ceilings gave the living room a spacious feel. An area rug partially covered a polished, wide-beam floor, and upholstered sofas and chairs added a comfortable and relaxed feel. It was a classic example of chic country casual, Napa Valley style. We decided not to look at property because there was nothing new to see. We were back to being Napa Valley tourists, enjoying our good fortune.

It was the real estate trip in February 1998, the second anniversary of when our search became serious, that told me our search might be winding down. I didn't feel the same excitement I had experienced earlier. My enthusiasm for traipsing across the country was beginning to wear thin. Perhaps it would be best to put my Napa Valley dream on the back burner. It couldn't be forced. We still liked the idea of having a second home, and since we had never found much to our liking in Florida, California locations dominated our thinking. We had purchased a week of time-share at Marriott's Desert Springs Villas in Palm Desert four years earlier and enjoyed our annual stays there. Perhaps California was the right place for a second home, but someplace other than the Napa Valley. With this new thought, Patricia and I planned a trip for early May that would take us to Indian Wells, Santa Barbara, and the Napa Valley.

Our first stop was The Vintage, an upscale development in Indian Wells near our time-share that we had first seen a year earlier. It was the most upscale development in the area, made more alluring to us by its name. The facilities were first class and extensive—two golf courses, a fine clubhouse, a health club, and well-manicured grounds. With a realtor, we toured three homes, one of which suited our tastes. But something was missing. The "something" was what to do with our time after we played golf and used the health club. We needed a worthwhile activity to engage our time and energy, and it was not apparent what that would be. We headed to Santa Barbara somewhat unsettled but knowing the desert held possibilities if we could figure out how to occupy our time.

Santa Barbara was experiencing heavy rains when we landed, and the weather didn't improve much during the next two days. The intermittent downpour and disappointing stay at an upscale, boutique hotel detracted from our enjoyment of an otherwise charming city. We wanted to look at homes on a golf course, but Santa Barbara offers few. The convenience of golf had become an important factor since we both enjoyed it. After looking at homes in the surrounding area, we took a trip to wine country over the mountains in the Santa Ynez Valley. Most of the wineries had been built in the preceding ten to fifteen years, classifying the region as still in its developmental years as a wine center. I felt a bit too old to be a pioneer, so we crossed off that area from our list and headed to the Napa Valley.

The entire California coast was now experiencing heavy rain, making our arrival in San Francisco as soggy as our departure from Santa Barbara had been. We checked in to the Vintage Inn in Yountville, had dinner, and decided to tour on our own the next day.

The morning was dreary, but just being in the Napa Valley gave us a boost. In the afternoon we looked at the three properties Steve had recommended. Patricia gave a thumbs-down on the first two. Our third and final stop was the house in Oakville I had visited with Steve the previous summer, and I gave her a description as we drove.

"This place originally came on the market about a year ago. It's on a ten-and-a-half-acre parcel with a walnut orchard and seven acres of vineyard. When I saw it last summer, I thought it was by far the best I'd seen, but I didn't mention it to you because it was too pricey. The owner let the listing expire in the fall, lowered the price to something we'd be willing to pay, and put it back on the market in January."

We stopped briefly at the corner of the property where the gravel driveway leading from the house met the road, but it was long enough for Patricia to say quietly, "This is it."

Her reaction took me by surprise. This was the first time we had both had an immediate positive reaction to a place. She hadn't been inside and there was nothing about the scene to prompt that comment. The house and its surroundings were unchanged from my last visit, and the dismal weather did nothing to improve their appearance.

"Why?" I inquired.

"It's the setting. And the house is the right size. You said it already has a vineyard," she said looking around. "It's off by itself but not remote."

"This is an A location," I said. "It's got everything we want." Pointing to the construction timbers and rock pile, I added, "The inside is a bit unkempt like the outside, but it has great bones. There's plenty of room for improvements, but it's a beauty."

"This is the size we want," she remarked, as a gray terrier with a faded red bandana around its neck strolled down the driveway to our car, gave us a cursory look, sniffed around the car, and sauntered back to the house. I told her the price.

"That's more than I expected. But it's much better than others we've seen," she said. "Let's get a hold of Steve and see if we can get in today."

We called him from a phone outside the Oakville Grocery about a mile and a half away, and he arranged for us to tour the property at 4 p.m. He met us there, and as we were exchanging greetings, the young woman I had met on my first visit came to the door. She apparently had been told to expect prospective buyers and waited for us. Excusing ourselves from Steve, we headed for the house. The drizzle of early afternoon had become a steady rain, and I noticed that water was pouring directly onto the ground from the roof, bypassing the gutter that had pulled away from the face board.

Once inside, the young woman invited us to look around and then headed upstairs to her room and closed the door. Patricia took her time to walk around the first and second floors, taking in both the good and bad features. The rooms were a comfortable size and were filled to overflowing with the owner's possessions. Among the more picturesque was an array of farm implements—hoes, saws, pitchforks—hanging on the walls of the family room. Outside, the serrated ridgeline of the mountains on the left and right and Mount St. Helena in the distance were visible through the haze. I waited on the covered porch as Patricia finished her tour and joined me. Looking toward the western boundary of the property, we noticed a small, rickety barn; a stable; and a corral about 150 feet from the house.

"All this place needs is a facelift," she said.

"I hope so. The gutter in the front says it hasn't gotten much maintenance. Makes me wonder what else needs fixing. Despite the drawbacks, it's the best we've seen. Let's make an offer."

We had both seen the quality of the structure and the value of the location. She would never see it under more adverse conditions, so there was something fundamental about the place that appealed to both of us.

As we approached Steve's van, he asked, "Patricia, what do you think?"

"I like it more than anything I've seen, Steve. We want to make an offer."

He cocked his head and looked at her questioningly. "That's a pretty quick decision."

"Pretty quick?" I said. "That's more than two years of experience talking!"

"I know that," he declared. "But you were in the house less than thirty minutes and know nothing about the vineyard, the condition of the septic system, or whether the house has ever been flooded. How do you know you want it?"

"When I think of all the places I've looked at, there's no doubt this tops the list," I responded. "The size of the house and the setting are just right. There's vineyard on all sides, so how bad can the soil be? If things in the house are broken, we can fix them."

He chuckled. "It's always good to know what makes your client want to buy."

A car came down the crushed-stone driveway and parked thirty feet from us in front of the house. A woman got out, waved in a manner that said, "Don't bother me," and went inside.

"I think that's the owner," Steve said.

With Steve in his van and Patricia and me standing under umbrellas, we discussed the offer. The house obviously needed fixing. We would have to make a major capital infusion to redo the master bathroom and bath, replace the gutters, and paint inside and out—and we still didn't know anything about the condition of the well, the plumbing, or the vineyard. As we talked, a second car came down the driveway. A man and woman got out.

"Hi, Steve," the woman said as she and her companion walked over to us.

"Hi, Pat," Steve responded. "I wasn't expecting you. Meet my clients."

We were introduced to Pat Taylor and Vinnie Ruggiero, realtors from Morgan Lane who had listed the property. They explained they were there for a 5 p.m. meeting with the owner that they had scheduled yesterday. Vinnie held up the bottle of wine in his hand and said jokingly, "It's obviously a very important meeting."

"Who's going to need most of that, Vinnie?" Steve inquired. "Don't know," he retorted. "We may all need some."

"Tell me something about the history here," Steve asked. "What other offers have you brought to her? What's it going to take to make a deal?"

"She didn't even counter to a couple of architects in Napa," Pat responded, quoting the figure they offered.

"Has there been much interest?" Steve persisted.

"In this market, Steve, there's always interest."

"I mean serious interest."

"Nothing at the moment," Pat replied. She added, "You've got to excuse us. We have to get inside to keep our appointment."

As she and Vinnie headed toward the house, she stopped

and turned back. "I suggest that if you have an offer, you get it to us as soon as possible."

"We were thinking of giving you one today," I said.

"We look forward to getting it," Pat stated as she continued up the walkway.

Rather than wonder if they were peering through the curtains at us, we reconvened our discussion in the Oakville Grocery parking lot. There in the May overcast, we prepared an offer that Steve agreed to deliver to Pat. It was now 6:15 p.m. He promised to contact us as soon as he got a reaction or a counteroffer. With nothing likely to happen before morning, Patricia and I headed for dinner to talk over what we had just done. A new restaurant, Bistro Jeanty, had recently opened in Yountville to wide acclaim, and we made it our destination. A mariachi band serenading an overflow crowd at a nearby Mexican restaurant reminded us that this was Cinco de Mayo and a celebration was underway. It was almost too corny to think we might have something to celebrate, too.

Things had moved fast. Patricia had caught me off guard with her quick "This is it" when we had first pulled up to the property. She hadn't reacted like that before and I couldn't tell whether her commitment to our offer was solid. Remembering Byrd Hill, there seemed a better-than-even chance she'd talk herself out of it over dinner.

Bistro Jeanty was a rising star in restaurant circles. Philippe Jeanty had distinguished himself for years as the chef at Domain Chandon and had decided to go out on his own, following in the footsteps of other successful chefs-turned-restaurateurs like Wolfgang Puck, Charlie Trotter, and Thomas Keller, whose French Laundry a short distance up the road had blossomed into one of America's premier dining establishments. We dined our way through tomato soup in a bread bowl, walnut salad, a bottle of Chardonnay, sorbet, and coffee. Not surprisingly, our conversation centered on our offer of a few hours earlier, wandering off into related subjects from time to time.

Throughout the meal, Patricia never wavered from her decision and reaffirmed it with all the reasons we had discussed, expanding on each as she spoke. During dessert, she made the observation, "You know, if we buy it and decide it's not for us, we can always put it back on the market. It doesn't have to be ours forever."

Her comment proved there could still be surprises in marriage even after thirty-four years. That point of view had always been my line. She typically reacted to every parcel we considered as though we would have to die with it if we ever bought it. It had become a sore point between us. Now she saw it differently. It seemed that we were going to see this deal through to the end.

Chapter 15:
Reaching Agreement

The next morning Steve called to relate the terms of the counteroffer that had just come in. It included only a nominal drop in the asking price. Whether we liked it or not, in her mind the owner had already reduced the price substantially from when she had first listed it a year earlier. We agreed to meet Steve in the parking lot of the Oakville Grocery, which had unexpectedly become the command center in our search for Napa Valley real estate. The store sells sandwiches, salads, and a wide selection of gourmet food and beverages and is a mecca for travelers to the valley. It does a booming business from the time it opens its doors in the morning to the local coffee crowd until the last tourist leaves in the evening. The building it shared with the post office amounted to "downtown" Oakville.

We revised our offer and Steve dropped it off at Pat Taylor's office. He called to tell us we would not be getting a response until later that afternoon. When it finally came, there was a change in the closing target date but no change in the asking price. At our powwow with Steve the next morning, we were debating the next step in the negotiating process when Patricia said, "After looking for more than two years, let's just accept the price and get it over with."

It was a reaction that was vintage Patricia. She has no tolerance for the give and take of negotiations and had tired of the process. Steve and I looked at one another, thought of all those property visits and drives up and down the valley, and realized we had heard words of wisdom. I took the counteroffer from Steve's hand, signed it, and gave it to Patricia, who added her signature. At 9:50 a.m. that morning, the deal was finalized. It ordinarily would have been a time for euphoria, but we had been this far before and still didn't own anything. Yogi Berra's observation that things are never over until they're over came to mind. Nonetheless, at least we had a contract.

Steve dropped off our acceptance at Pat's office and returned to his office on Main Street in St. Helena. There he, his wife Marla, Patricia, and I met to prepare the list of actions that had to be taken, focusing initially on the inspections.

"Did Pat have any reaction?" I asked when Steve arrived.

"Yeah. She was a little surprised. I don't think she expected you to move that fast."

"Did you find out anything about that meeting she and Vinnie had with the owner the other night?" Patricia inquired.

"I asked but she just smiled. Said she'd tell me one day, but not now."

Steve's company had one of the smallest and least-pretentious real estate offices in the valley. The nationwide brokers like Coldwell and most local brokers had larger offices, and Patricia and I had expressed concern to each other about the ability of Steve and Marla to follow through on all that had to be done after the contract was signed. We quickly found our concerns were unwarranted. Working with them was a pleasure. It became apparent they were an experienced team. Whether by design or by aptitude, Steve was "sales" and Marla was "service." They understood better than we what had to be done and they knew the best people to do it. We quickly agreed on the need for a thorough inspection of the house, an evaluation of the vineyard, a check of the septic system, and an analysis of the well.

On our flight home, the full scope of the adventure we were beginning sank in. To make this work, we had to depend on people we had yet to select: an architect, a home-remodeling contractor, an interior decorator, a manager for the vineyard, a gardener, and someone to look after the house when we weren't there. We debated whether we should build a second structure on the premises for someone who would take care of things when we wouldn't be there. The scale of the change we were making in our lives was increasingly apparent. What had been over the horizon during the past few years was now imminent.

The inspections went smoothly. The credit for that went to Marla, who followed up and followed through on every detail. One of the earliest reports came from a local laboratory that had tested the water sample from the well.

The water was safe for drinking but was high in iron, manganese, and boron. In fact, the amounts of manganese and iron were five times greater than was needed to stain porcelain. That explained the discoloration in the sinks and toilet bowls.

The boron was another story. It was at an acceptable level from the standpoint of human consumption, but it was much higher than recommended for vineyard irrigation. The lab technician explained to Marla that the amount of boron in the water was bad for the vines. I wrestled with what to do. Steve assured me that water treatment would easily remove the excess manganese and iron. Unfortunately, nothing similar was available for boron as far as he knew. It sounded like a problem, but I didn't have the experience or know-how to determine its significance. In addition, we still needed other reports to come in before having to decide whether or not to follow through on the purchase.

The vineyard manager I had met through Steve agreed to give us an opinion on the vineyard, which was divided into two blocks. Closest to the house were three acres of Sauvignon Blanc, and he indicated they looked healthy. However, he noted that the four acres of Chenin Blanc behind it were in bad shape. He doubted whether there would be much of a crop and suggested it be pulled out at the end of the season. The filters on the well would have to be replaced, and the discoloration around the pipe joints substantiated the fact that the water was high in iron. The good news was that the soil looked very good.

A few days later, while Steve was giving me a favorable update on one of the inspections, I commented that I did not remember seeing an air-conditioning unit outside the house and wondered if I had missed it.

"Come to think of it, I don't remember if there is one," he remarked. "I'll be going over there tomorrow afternoon, so I'll look around and give you a call."

He called later that day. "You got me wondering, so I went over right after we spoke. There is no air conditioner. Since the owner was in the house, I asked her about the heating system." He paused. "I don't want to shock you, but there is none unless you consider a central heating system to be a fan that blows hot air from the fireplace through some ductwork."

I had never thought to look. The heat registers in the floor and a propane tank right behind the house suggested there was a heating system. Could a fireplace really keep it warm in the winter?

"I know you wouldn't kid me on something like this," I muttered.

"Believe me. I'm not kidding. There still are houses around here without central heating but they tend to be smaller and older. Must get nippy in there in December and January." It was apparent that our project list would have to be expanded to include central heating and air conditioning.

The reports confirmed that the house was well built. Although the shower pan in the master bath was cracked, allowing water to seep into the carpet, and a water pipe in the wall leaked, our plans to remodel would fix them. The only item of concern was a crack in the cover of the septic tank. Regrettably, the deck had been built out over it, making it impossible for the inspector to determine if the system itself had been damaged.

During one of my regular calls with Steve toward the end of May, he said, "I was talking to Vinnie today and asked why he and Pat went over to see the owner that day we met them. He laughed and said the owner had decided the previous day not to sell. When she called to tell Pat and Vinnie, they asked to meet with her because there were still two months to go on the listing contract. That's what they were there for. Apparently she and her boyfriend had decided to take it off the market. She didn't tell him about the contract with you until she signed it. I guess they had a blowup. Now she's sounding to Pat and Vinnie like she wants to get out of the deal."

"Can she?" I asked.

"No. The contract is binding. The realtors produced a qualified buyer, got a contract signed, and earned our commissions. I can't imagine that Morgan Lane would be willing to give up their piece. I know we're not willing to walk away from ours."

"Even if you did, what are my rights?"

"You can sue for specific performance."

"Should we have an attorney look at the contract?"

"I don't think you'll have to. There will be enough time for that if it gets messy."

If the owner had any doubts about wanting to sell, the speed with which we came to an agreement gave no hint of them. It looked to us like all systems were "Go."

Ehrlich Vineyard Photos

The Oakville Grocery represents "downtown" in Oakville, a community in the heart of Napa's Cabernet Country.

These Cabernet Sauvignon clusters have come through a good growing season and are ready for harvest.

Sauvignon Blanc enjoys the soil and climate of Oakville.

Brian and Christina's children, Emilia, Bella, and Avery (left to right),
share the spotlight with newly harvested Sauvignon Blanc.

Half-ton bins of Sauvignon Blanc are checked
before their trip to the Groth Winery.

Mustard is included in the ground cover that gets planted in the
vineyard for winter. It adds organic nutrients to the soil when it is
disced under in the spring.

Cliff, Robert Mondavi, Patricia, Margrit Mondavi, and
Susan Ehrlich are all smiles at the dinner honoring Robert held at
the Groth Winery in September 2001.

Robert Mondavi (center) holds the poem about him that Cliff
composed and read at the September 2001 dinner. It was presented
to him in his office by Cliff and Patricia.

Cliff, Patricia, and Michael Weis, the Groth Winery's winemaker
since 1994, sip some Sauvignon Blanc juice immediately after the
grapes are pressed in 2005.

Robert Mondavi (left) and Dennis Groth (right) chat during Robert's
90th birthday celebration at the Opus One Winery in June 2003.

Neighbors Toni and Frank D'Ambrosio helped the
Ehrlichs settle into the Napa Valley.

It was Steve Ericson (right) who steered Cliff (left) and Patricia
(second from right) to their vineyard in Oakville. His wife, Marla
(second from left), made the closing run smoothly.

Filiberto Chavez and Becke Oberschulte are
the landscaping "A Team."

Guy Larsen relaxes at the Tenth Anniversary Party
with his wife Maureen.

Cliff holds a bottle of vintage Freemark Abbey Cabernet Sauvignon from Laurie Wood (left) who witched the Ehrlichs' new well.

Tom Pulliam drilled the well that solved the Ehrlichs' water problem.

Cliff and vineyard manager Paul Garvey discuss
grape farming operations.

Nils Venge (owner of Saddleback Cellars), Christina Ehrlich, and
Brian Ehrlich enjoy the Tenth Anniversary Party.

Susan, Patricia, Scott, Brian, Christina (Brian's wife), and Cliff are
ready to pour at the Tenth Anniversary Party.

Brian and Christina Ehrlich's daughters – Bella, Emilia, and Avery –
ham it up for Cliff's friend Larry Guest during a visit to the vineyard.

Jeff Fontanella gets ready to sample some
2012 Ehrlich Cabernet that's been aging in oak.

Neatly pruned walnut trees share the land with rows of grapes.

B Cellars owners Jim Borsack (center) and Duffy Keys (right)
tour their vineyard with winemaker Kirk Venge (left) soon after the
winery opened in September 2014.

Cliff Ehrlich, Joel Alper, Patricia Ehrlich, and Louise Alper (left to
right) are having no trouble enjoying an afternoon
in the B Cellars Tasting Room.

Cliff and Patricia practice the fine art of "dropping fruit" to help achieve a canopy-cluster balance of their Cabernet vines.

The crew that bottled the 2013 Ehrlich Cabernet in
December 2015 poses for a photo.

B Cellars uses a distinctive screening process for their wine bottles
where the lettering is first painted on and then baked using a kiln.

Tasting party on July 27, 1968, in Anaheim, California.

Photos by Patricia Ehrlich, Susan Ehrlich, and Larry Guest.

Chapter 16:
The Deal Gets Closed!

None of the inspection reports that rolled in gave cause for serious concern. The story about the owner wanting to back out of the contract was never substantiated. I learned that she acquired the land from her father and lived on it in a trailer before building the house in 1984. Her mother still resided in a house in the grove of trees about half a mile north of the property.

It is, of course, at the closing that you take title to property you buy. However, the closing procedure we had experienced so often in the past—of buyer and seller meeting with lawyers to sign documents—apparently has become a relic of the past in California. It still may occur in some situations, but not ours. Patricia and I followed the instructions contained in the thick escrow package sent to us in Maryland and returned it to the title company. Included in the package had been a two-page document that made both of us smile. To be sure we knew what we were getting into, the county required us to verify that we understood that living in an Ag Preserve would expose us to "inconveniences or discomforts arising from noise, odors, dust, chemicals, smoke, insects, and the operation of machinery during any twenty-four-hour period" and that we "should be prepared to accept such inconveniences and discomforts as a normal and necessary aspect of living in a county with a strong rural character and a healthy agricultural sector."

We were ready. Bring it on.

On Tuesday, June 16, 1998, we became the owners of a

house and vineyard on 10.56 acres in the Napa Valley. We had a quiet dinner at home in Bethesda and toasted our good fortune with a bottle of the first vintage from Vineyard 29. The details of the search were etched in our brains, but we had no clue what our lives would be like once we owned a house and vineyard 2,500 miles away. That adventure still lay ahead.

Although the deal was done, I couldn't help but second-guess myself and wonder if we had paid too much. I was having a minor case of buyer's remorse until I ran across a quote from Rupert Murdock that appeared in *Business Week*. Being criticized for overpaying to make an acquisition he thought was critical to his company, Murdock said, "Sometimes you have to pay a high price for opportunity." Murdock's comment put my mind at ease.

I planned to move in as soon as possible, and Patricia would follow me in late July. I arrived at the house the day before the seller was scheduled to leave and found her busily packing and cleaning. She was using the move as an opportunity to get rid of much of what she had accumulated over the previous fourteen years.

I dropped in on Steve. In sports terms, he had become my "go to" guy. I was relying on him for more and more of the follow-up tasks related to taking occupancy and getting the house into shape. He had a wealth of information and a network of relationships that were invaluable to us. We needed a security system, a locksmith, a vineyard manager, a remodeling contractor, an interior decorator, a store to purchase appliances and phones, someone to help me evaluate my water situation, and someone to help me unravel county building codes to determine whether we could build a second dwelling for an onsite manager. He would listen, make a note or two, and then give me some names.

We held title to this real estate but allowed the seller to live there for two weeks after the closing. That meant we remained on the outside looking in. It was the arrival of a Macy's delivery truck that represented our first assertion of ownership. As promised, Macy's was delivering six mattresses, box springs, and bed frames Patricia and I had ordered earlier in the month. The driver and helper unloaded them onto the porch, and I wrapped them with high-visibility tape so they wouldn't get confused with the outgoing furniture. Their presence said, "The new owners are moving in!"

I was in my car in the driveway talking to Steve on my cell phone the next afternoon when the seller walked out the front door, closed it behind her, paused, and then hurried down the steps. She hesitated a moment when she saw me, but got into her car and pulled away. This was her final departure from a house she had built fourteen years earlier. Even across the driveway, I could see that her eyes were red-rimmed. It was not the time for social niceties. I just waved.

Steve arrived unexpectedly and drove slowly down the driveway. He had a relaxed expression.

"It's amazing how different this place feels to me now that it's yours," he said, shaking his head. "Just amazing."

He was right. It did feel different. You own a home emotionally only when the prior owner leaves and it's empty when you walk in. A preoccupation with my project list almost caused me to miss the experience. It was time to revel in the moment that had been two and a half years in coming. This was the finish line, and passing it went almost unnoticed.

"You're right, Steve," I observed. "It feels different because it is different. What's different is that all the problems that belonged to her now belong to me."

"I'm glad you aren't so emotional that you lost your sense of humor," Steve said.

"I'll probably need it to get through the next few weeks."

Chapter 17:
Water, Water ... Anywhere?

I had searched for two and a half years and looked at more than eighty pieces of property before Patricia and I found one we both really liked. It was the right size in the right place. The house needed some upgrading but was well built. The water situation was the only loose end. What had caused me to minimize its importance was the fact that thousands of real estate transactions occur every day and rarely do any of them hit a snag over water. I had pushed into the background the reality that most residential homeowners rely on municipal systems and know they can turn on the tap and not worry about what comes out. However, water availability and quality are threshold concerns in a rural area with no such system.

Patricia and I hoped the boron reading was an aberration and speculated that another water source could be found on the property. After all, we had ten and a half acres. Our emotions had yo-yoed up and down the way they usually did when we were considering a big-ticket purchase, but we had eventually decided to go to closing without a solution to the water problem. It was a case of ready, fire, aim and we knew it could backfire. If no solution to the water problem could be found, we would own the most expensive hayfield or organic vegetable garden in the Napa Valley.

One certainty was that the well needed a major overhaul. The pump was twelve years old and the filters leaked so badly while the vineyard was being irrigated that the spray from corroded fittings soaked anyone standing nearby.

It needed major surgery, but that wasn't a smart way to spend money if a new source of water was needed. Some research was required.

Fortunately when vineyard owners in Napa are in a jam, there are helpful resources available at the county's Department of Agriculture and the University of California's Cooperative Extension office. Ed Weber, the farm advisor at the extension office, listened to the explanation of my situation. At the mention of the boron level, he interrupted.

"You have a problem," he announced. "That's well above the recommended maximum. Boron is toxic to vines above that level and you're quite a way past it. How sure are you of the results?"

While the test had been made by a reliable laboratory, Ed observed that the sample may have been bad and suggested testing again. We did so the next day, but the report we got back showed virtually the same results. It was looking more and more like the solution would have to be a new well. That was when Steve suggested contacting either a geologist or Laurie Wood, the valley's premier water witcher who enjoyed an excellent reputation for finding water. Other people offered the same advice. No one who made the suggestion could name a geologist, but all of them had Laurie's name on the tip of their tongues. That made the decision easy. I called Laurie, left a message on his answering machine, and continued to get educated about wells and water.

Two days after we moved in, the sound of tires on our crushed-stone driveway heralded the arrival of our first visitor. Alighting from a dark-blue Chevy Suburban was a compact man in jeans, a short-sleeved shirt, and sunglasses. He stood with his hands on his hips and surveyed the house; the colorful leaves of variegated viburnum; the red, pink, white, and yellow blooms on the rose bushes; and the small front lawn. He seemed hesitant to approach the home of people he had never met, hoping his presence would soon be noticed.

"Good morning," I said, walking from the house.

"Yes it is," came the reply. "And that's true of most morn-ings. It's probably one of the reasons you're here."

The unexpected guest introduced himself as Frank D'Ambrosio. He said he lived in the house to the west of us, across nearly a thousand feet of vineyard. Patricia joined us as Frank explained he had grown up in Palo Alto and had come to the valley with his wife Toni a dozen years earlier. Four years ago they had moved into their nearby home. Learning that owning a vineyard was a new experience for us, he re-marked, "I bought a vineyard shortly after I came to the valley. I didn't know anything about it and no one helped me. I had to learn on my own. It was tough. I'd like to save you from some of the mistakes I made."

It was a gesture that fit Frank's friendly and relaxed per-sonality. He remarked that few properties ever went on the market in this area of the valley, and said he was surprised this one had been on the market for so long. We spent a few minutes getting acquainted, and then I told Frank my concern about our water situation. He listened as I explained our predicament and then commented very matter-of-factly, "The prior owner was aware of it. She thought it would make it tough to find a buyer."

His words put a knot in my stomach. It was one thing to know that a water analysis identified some kind of mineral problem. It was entirely something else to think that someone who had lived her entire life in the Napa Valley considered the situation so serious that she had decided to sell. Had our zeal to find a vineyard after a long search caused us to see an ugly duckling as a beautiful swan? Frank's remark was not reassur-ing. With errands to run, he repeated his willingness to help, promised to drop over during the week, and got on his way.

While I waited to hear from Laurie Wood, Steve made

other contacts, trying to shed some light on the problem. The owner of the large adjacent vineyard said his water had never been tested but gave no indication it was high in boron. A call to a water-system service company disclosed that it was possible to extract boron from water, but the process was expensive and the material you'd get would be a toxic waste that had to be disposed of properly. That was no solution. From one of his contacts, Steve learned of a retired petroleum geologist who owned property that was about as far west of the Napa River as our new home was east of it, and we contacted him. He invited us over.

His single-story house on the valley floor was protected from the sun by a stand of thirty-foot-tall fir trees. Its paint was faded and peeling from years of inattention. A tall, lanky man in his seventies, the geologist opened the door after a couple of hard raps got his attention. He had thinning hair and wore jeans and a loose-fitting shirt.

"Sorry," he apologized. "I was out back and had forgotten you were coming over. Come in. I'm Morrow."

Morrow lived alone in a house that had been furnished in an earlier time and did not appear to host many visitors. The entry hall was bright but the blinds in the adjacent rooms were drawn. The absence of sunlight and fresh air kept them cool and musty. Books and magazines stacked on tables and shelves suggested that Morrow was an avid reader. We followed him to the kitchen where he leaned against the sink, sipped beer from a can, and asked me to explain my dilemma.

When I finished, he said, "I was a geologist before I retired, so I have ideas about finding water that are not just theory. They may not be worth a damn, but I'll share 'em with you anyway."

Morrow had spent his career working for oil companies. He had roamed the globe for them, wrestling with challenges like extreme heat and cold, nasty insects, unruly natives, and surly governments. We were a new audience for him, and he related his experiences to us with such animation and gusto that it was easy to forget they had occurred years earlier. Eventually his reminiscences brought him back to the present.

"Geologists know that rivers meander over time and leave beds of fine gravel in the areas they pass over," he said as he traced a river on the countertop with his finger. "My theory is that you'll get your best water by drilling as close to the current riverbed as possible." He emphasized his point by tapping his index finger on the bank of the imaginary river he had drawn. "Drill until you're just past the first layer of gravel. You'll know it when you see it. It's small, round pebbles." He laughed a big guffaw and added, "Even if you don't know it, your driller will recognize it unless he's an idiot. By the way, who are you going to use?"

"The Pulliams," I volunteered, referring to two brothers Steve said were the most reliable drillers in the valley.

"They have a good reputation. They'll know what to look for. To get sweet water near the river, you don't want to go deep. Go shallow. My best well is only sixty feet deep." He paused and repeated himself. "You don't need to go deep for good water."

His theory sounded fine, but I had no basis for evaluating what he said. At least he had provided some reference points and a plausible explanation of what was needed to find "sweet water."

With everyone I spoke to contributing to my education, I contacted the drilling company that competed with Pulliam and met with the owner's son, a lean, leathery-skinned man in his mid-thirties. He listened to my story with his thumbs hooked in his belt, shifting his weight from one foot to the other and occasionally glancing down at the ground. I had barely finished when he jumped in to explain that his company charged by the foot and preferred to drill deep—really deep—

rather than shallow. When I inquired whether it was best to drill shallow or deep to avoid boron, he shrugged and said, "There's a lot of theories. I just don't know what to tell you."

It was an unexpected response, since his family had been drilling locally for more than a quarter of a century. He moved impatiently, realizing instinctively that his unwillingness to help had brought the conversation to an end. I thanked him and we shook hands politely, the way people do when they know it's unlikely they will ever see one another again.

The next morning, I answered the phone and found myself talking to Laurie Wood.

"I got back in town yesterday and have messages from you, Steve Ericson, and Bill Pulliam. Must be an emergency," he said.

"It is to me," I responded and then gave him the details.

"I'll be out there in a day or two," he volunteered.

The waiting and the uncertainty of how serious our problem was had made me increasingly anxious. It was the kind of anxiety that comes from knowing you eventually have to make a major decision on a subject you know very little about, based on advice given by people you have never met before and who have no stake in the outcome. It wouldn't take Laurie long to figure out I had no experience on a subject he understood in great detail. To him, I was just another guy—a newcomer no less—with a well problem that he probably had seen dozens of times before. By the time Laurie's blue pickup rolled to a stop in my driveway for his first visit, my discomfort level was soaring. I was having a serious Maalox moment.

Chapter 18:
Relying on the Drillers

It was reassuring to finally meet Laurie, tour my vineyard with him, and learn something about his skill at finding aquifers. His helpfulness improved my frame of mind and flushed the angst from my system. He was a coach by instinct. In just over an hour during our first meeting, he had shared information about insect problems associated with our walnut trees, explained how to recognize the signs of boron toxicity in vines, provided reassurance about the durability of my vines, explained why the Chenin Blanc didn't have irrigation, conducted a short course on vineyard diseases, discussed the advisability of frost protection, and given some hints on what could be done to salvage this year's crop.

He was back a week after his first visit, explaining that he preferred to work by himself, and then walking the property with his rods, trying to find water. I watched him for a few minutes from the window before deciding to let him alone. After an hour, he returned to the house.

"I don't know what to make of the boron situation," he said. "I've never faced it in this area and don't know what we'll find, but I marked two sites. One is out in the vineyard just south of that lone walnut tree. If you drill down about eighty-five feet, you should get at least a hundred gallons per minute. The other one is near the southwest corner of the property. Down about ninety-five to a hundred feet, you should get at least a hundred gallons a minute. They should give you enough for both vineyards and, if we're lucky, maybe enough for frost protection, too."

The alternative, he pointed out, was to drill deep and go past the boron. The problem was that it was impossible to tell how far was far enough. The conventional wisdom favoring deep wells had changed over the years, and now shallow was thought to be better.

Sitting in the shade on the back deck overlooking the Sauvignon Blanc, I asked him if his roots were in the Napa Valley. "I was born in St. Helena in 1920," he volunteered. "June fifth, to be exact. My father owned thirty acres of walnut and prune orchards, so I grew up on a farm. Right after school ended every year, the harvests would begin, and they would continue right through the summer. All of us would be picking. Peaches ripened around the first of July and pears in the middle of the month. Prunes were ready in mid-August and then it was grapes and walnuts."

Laurie served in the army during World War II and was a bona fide member of our nation's Greatest Generation. His rural background and familiarity with farm equipment got him assigned to a road-construction group, and operating bulldozers became his specialty. He worked on projects like the Alcan Highway until 1944, when he was shipped over to England to participate in the invasion of Europe, landing in Normandy on June 9, 1944. As a combat engineer, he fought at the Battle of the Bulge. After the war, he returned home to join his father's new vineyard and orchard-management company and took it over years later when his father passed away. He spent most of his time running it and found water as a sideline.

"Water witching and water dowsing are terms that are used interchangeably," Laurie observed. "I got involved with it in 1950. A couple of old-timers, friends of mine, had a prune orchard on Zinfandel Lane. I saw them out in their field one day with their copper rods and wondered what they were doing. They must have had a couple of snorts at lunch 'cause they were slapping each other on the back and having a grand old time. They explained how the presence of water could be sensed through the rods if you concentrated, and then gave me some to try it out. I walked around with them and found I had a knack for it. Each time I saw them over the next few months, they taught me more about it. I got pretty good, but for years Elwood Mee did most of the dowsing in the valley. People learned about me gradually, and after Elwood died I got most of the calls."

I was interested in Laurie's story and about the idea of water witching, but I was skeptical too. Water witching certainly has its detractors. A geologist writing in a local magazine had called it a superstitious practice based on ignorance, adding that anyone who believed in water witching was capable of believing the world was flat. Maybe so, but Laurie's reputation for tapping into the aquifers flowing beneath the ground couldn't be ignored. There were hundreds of people in the valley who started the day washing their faces with water from sources Laurie had located, and dozens of vineyards were irrigated by wells he had found.

A few hours after he left, Bill Pulliam called to say that his brother Tom would be coming the next day to set up his drilling rig. It was close to noon when Tom arrived with his son Aaron. They were a pair of hearty and robust souls, well suited for the rugged work of well drilling. Tom had a salt-and-pepper beard, graying hair, and an easy smile. Aaron was a lankier version of his dad and sported a full, droopy handlebar mustache that gave him the appearance of an actor in a western movie. They looked like the proverbial pair that would beat three of a kind.

"You're gettin' us at a good time," Tom growled. "We just finished up a well in Yountville that came in at three hundred gallons per minute, and last week we drilled a gusher nearby that's eight hundred gallons a minute." He added with a grin, "We don't want any dusters here to break our hot streak."

Their drilling derrick lay on a flatbed truck that Tom positioned at the walnut tree site in the vineyard. While he and Aaron were getting it into place, Laurie came by in his pickup.

"I owe you one, Laurie," Tom said, pointing to the tree. "Can't remember the last time you found a site that had shade."

"If I'd known it was going to be you, I'd a found a different one and made sure you baked all day," came the retort.

Tom and Laurie were old friends, and as they exchanged stories about their drilling experiences, it became apparent that my situation was not uncommon. They had seen variations of it a hundred times over the years. It was their regular fare. Tom had been a well driller all his life and said the drilling he did on the valley floor usually had a happy ending. No guarantees, of course, but there were more successes than disappointments.

"If Laurie got those rods of his crossed and we don't find water at this spot," Tom snorted, "we'll just keep punching holes till we do."

The Pulliams got underway the next morning. Operating his rig from a control panel at its rear, Tom began boring into the soil with a bulbous mud drill almost a foot in diameter. The material it cut through was floated up to the surface by water pumped into the hole from a tank on the truck. The slurry discharged onto a mesh conveyer that deposited the solid material on the ground near Aaron. The combined noise from the drill, the pump, and the conveyor made conversation impossible, and Tom gave me a set of the same kind of earplugs he and Aaron used.

When only a few feet of each twenty-foot drill stem remained above ground, Tom hydraulically maneuvered another into place above it, threaded the two together, and continued drilling. It was only an hour after they fired up that Aaron beckoned for Tom. They huddled as Tom took some of the material off the conveyer and rubbed it between his fingers. He nodded to Aaron, walked over to me, and motioned for me to remove one of my earplugs. Above the din, he said, "This is gorgeous material. It should be carrying a lot of water."

The fine sand and smooth pebbles in his hand resembled material from a riverbed. Morrow was right. It was easy to recognize.

"We're down at seventy-four feet now," Tom continued. "We'll keep drilling till we get through this strata."

It was another ten feet before the machine once again pumped clay from the hole, and Tom stopped drilling. With the hole complete, Tom removed the drill stems and returned them to the storage area on the rig. He then fed a twenty-foot-long piece of six-inch-diameter PVC pipe into the hole. The pipe was capped at the bottom and perforated with a series of narrow three-inch slits. The top was flared to allow the next piece to fit snugly into it. A coating of glue and a screw were used to make a tight seal. Tom added other pieces of pipe and sawed the final one off about three feet above the ground.

Aaron shoveled pea gravel in to fill the space between the casing and the side of the hole, but filled the final twenty feet with concrete to provide a sanitary seal after the county inspector gave her approval. Before capping it off, Tom tested the flow rate by injecting a stream of pressurized air and determining how long it took the stream of water that spewed out to fill a thirty-two-gallon trash can. Four separate tests averaged thirteen seconds each. The well was producing almost 150 gallons per minute. Not a gusher like Tom and Aaron had found nearby, but plenty for our needs. After Tom drew a water sample for testing, he and Aaron began to break down their equipment to move to the second site.

About that time, Julio Delgado arrived to finalize plans for building a stone wall across the front of the property. Julio had done work for the D'Ambrosios, and Frank had given him an unqualified recommendation. The wall was to be 250 feet long, 24 inches wide, and 30 inches high, with each end anchored by a pilaster. It would make use of one of the natural resources on the property: a pile of rocks three feet high and eight feet wide that ran a distance of one hundred feet. The previous owner had acquired them when a nearby winery was excavating to make storage caves. She had planned to build a stone barn but never saw it through. Patricia and I decided that rock walls were a picturesque feature of Napa Valley, and our rock pile could be used to build one. With the water picture brightening and construction of the wall about to get underway, prospects for our vineyard were definitely improving.

Chapter 19:
Plenty of Projects and New Friends

I had retired to be more independent, and now I was experiencing what that meant. An ever-growing list of projects monopolized my attention, and I quickly realized I was where the buck stopped.

The projects fell into a few natural categories—interior remodeling, exterior improvements, landscaping, the orchard, the vineyard, and of course the water system. Despite what was accomplished every day, the number of unfinished tasks blinded me to the progress being made. In fact, during some days the "to do" list grew longer. That occurred most often when the completion of one task led to a series of other tasks that had been overlooked or given only a passing thought. The remodeling of the bathroom and closet was an example.

The remodeling work was contracted to Guy Larsen, a builder I met through Steve. He knew from architect Don Clark's construction plans what had to be done and ordered all the material—studs, wallboard, joint tape, and the like. I played no role in those decisions. Once the framing and wallboard were in place and the rough work neared completion, the selection of fixtures, tile work, mirrors, and lighting moved to the forefront. A brief entry on my project list—select bathroom fixtures, tile, mirrors, and lighting—masked time-consuming activity. Understanding the cost and quality trade-offs, coordinating the colors, making the selections, and coming to an agreement with Patricia all took time, as did the inevitable minor tweaks to fit onsite realities that were beyond the scope of the architectural drawings.

Obviously, not all projects were equally demanding. Getting the house re-keyed and buying appliances were far less complex than selecting bathroom fixtures and correcting the water problem. Each project took time. For some, however, the challenge was not time but the emotional load it carried. The project that christened me to agricultural living fell into that category and occurred shortly after moving in.

I had seen what were euphemistically called "droppings" in the small room behind the kitchen, which housed the water heater, washer, dryer, and storage cabinets. A call to the local pest-control company brought a technician who confirmed my diagnosis. He put poisoned bait behind the water heater, under the sink, and in the crawl space under the house.

"This should take care of those critters in a week or two," he explained. In a week or two. Was he kidding? I wanted them finished off now. Today. Pronto.

"Can't we do it any quicker?" I asked.

"How?" he asked. "You don't know how many you have or where they are hiding. All you can do is give the bait time to work."

Fortunately, a solution to the critter problem presented itself within a week, but it came from an unexpected source. Guy Larsen took it on as a project and discovered that the critters had burrowed tunnels under the foundation to get into the crawl space beneath the house and then into the living areas. Once he and his crew plugged them, the problem vanished. When you need to solve a farm problem, I discovered, you need to find someone who grew up on a farm.

It didn't take long to discover that Guy was like Laurie. A lean, wiry, thirty-something who had grown up in the valley, he had as fair a complexion as you would expect a Larsen to have. His father was a woodcutter who had taught his son his trade. Guy, however, gravitated toward construction, developing a skill in carpentry and then starting his own construction and remodeling company. While he was working for me, any problem I had was his problem. He pitched in whenever he could. The horror stories I'd heard about remodeling projects running amuck were foreign to my experience with Guy. He was a meticulous craftsman as well as a cost-conscious contractor.

The other person who helped lighten the project load was Becke Oberschulte. Within an hour of Frank D'Ambrosio's visit in June, two people wearing tennis outfits and riding bicycles showed up in our driveway. A melodious "Hello in there!" brought Patricia to the door, and she was already talking to our visitors by the time I joined them. They were Clarke and Elizabeth Swanson, who lived on the south side of the access road to our property.

"Welcome to Oakville," Elizabeth enthused. "It's such a lovely place. Clarke and I wanted to introduce ourselves and see if there is anything we can do to help you move in."

Elizabeth enthused naturally. Tall and trim, she was an outgoing and upbeat person whose effervescence surpassed anything a Welcome Wagon could offer.

"We had heard the property had changed hands, but weren't sure when the new owners would be arriving until we saw you last week," she continued. "You bought a wonderful home."

"There's always more to moving in than you anticipate, and we'd like to be of some assistance if we can," Clarke volunteered. A few years senior to Elizabeth, he had flowing gray hair and a dapper appearance. It was reassuring to discover that both he and Elizabeth wanted to lend us a hand in getting settled.

"We don't expect to spend the full year here and need someone to look after the landscaping and property for us when we're away," Patricia explained. "If you know of anyone who could help us, we'd appreciate it."

"You should speak to Becke," Elizabeth said. "She looks after our landscaping and is very reliable. She's on vacation at the moment but is due back next week. She may be interested or she will know someone who is."

Elizabeth and Clarke helped make us feel welcome. A few days later, I imagine they were pleased when we hauled away two truckloads of timbers and assorted odds and ends that had been spread across the front of our property, which they had to look at every time they used their driveway. Their advice about Becke could not have been more on the mark. She had a thriving business attending to the landscaping of half a dozen clients. A charming woman with an easy smile and warm personality, Becke enjoyed being among flowers and plants. She was doing what she loved and it showed. She, Patricia, and I walked around the property discussing what had to be done and getting to know one another. She pointed out a couple of items we had overlooked and clearly demonstrated she had the know-how to be of great help. She was cut from the same cloth as Laurie and Guy. Between Steve Ericson, Laurie Wood, the Pulliams, Guy Larsen, and Becke, we had an A-Team that provided the support we needed to get up and running.

As soon as the Pulliams finished drilling two wells, we tested the water from each but discovered that the boron levels were only slightly better than at the existing well. I conferred with Laurie, who was disappointed, and he suggested that he find another site. I agreed, but the Pulliams had drilling commitments for the remaining non-rainy months of the year and would not be available until the next spring. The water problem was still the five-hundred-pound gorilla in our lives, but for the time being, it would have to be put on the back burner.

My focus shifted to salvaging the Chenin Blanc, but that wouldn't matter if we couldn't sell it. With no contacts in the wine industry and not knowing exactly how to proceed, it was necessary to improvise. The shelf with Chenin Blanc in the wine section at one of the Vallergas Markets in Napa provided the names of some local wineries that produced it. There were only three, and I called each one to see if any of them were interested in purchasing my grapes. George Buonaccorsi at Beringer was the most interested. To avoid any misunderstanding, however, he made it clear that Beringer had more interest in developing a relationship with someone with seven prime acres on the valley floor than in buying this year's harvest. Our site was a good one, and he anticipated that whatever varieties we grew would be the quality Beringer would like to buy. We signed the contract he proposed, hoping that the income from eight to ten tons would provide a much-needed cash infusion to offset some of the expenses of our remodeling projects.

We installed a new septic tank rather than worry whether the cracked cover on the existing one was the forerunner of more serious problems. For the same reason, we replaced the irrigation system used for the lawn and landscaping rather than just repair it. The walnut trees that formed a sagging canopy over the driveway were limbed up. The three whose branches hung over portions of the house and made the roof a raceway for squirrels were substantially pruned. Squirrels had even chewed a hole through the siding near one of the attic vents and gained access to the attic. Guy evicted them before replacing the damaged board with a new one. Julio Delgado and his cousins from Mexico were making steady progress on the rock wall that would stretch across the front of the property.

Patricia and I gulped one day when we realized that twenty-one people were working for us on one job or another. There were Guy's carpenters, Julio's wall builders, a heating and air conditioning crew, a paint crew, a crew to replace the gutters and downspouts, a tile man, an electrician, and a plumber. We felt like the Napa Valley Economic Improvement Commission. In talking about the early years of Vineyard 29, Tom Paine had remarked that he seemed to be writing checks all the time.

I was having the same experience. Apparently it's part of the vineyard initiation process.

In the meantime, we had begun to enjoy our new surroundings. The installers were literally tacking down their last piece of carpet when our first carload of guests drove down the driveway. Our daughter Susan had decided to celebrate her birthday with a party at our new house. Patricia and I had kept our children abreast of our search, and they all shared in the excitement when it came to a successful conclusion. Susan lived in New York, but she had many business-school friends in the Bay Area and invited them to Oakville. With a limited amount of furniture but with all bathrooms operating, we hosted our first guests two months after moving in. They popped a cork on a Schramsberg sparkling wine during the evening to toast Susan's birthday and our new house. The clinking of glasses was a symbolic announcement of our arrival in the Napa Valley.

Chapter 20:
The Odd Couple

When it came to teaching me the vineyard business, Frank D'Ambrosio, my new neighbor, proved to be as good as his word. He followed up his first visit by coming over regularly to discuss grape growing. In addition to the fourteen acres of Merlot that surrounded his house, he also grew Chardonnay and Merlot on a vineyard in Napa. Frank had learned the vineyard business from the bottom up and understood what was required to grow grapes profitably.

Part of his tutelage included my participation in discussions he was having with prospective vineyard managers. He was dissatisfied with his current manager and had decided to interview others. He invited me to accompany him on a walk through his vineyard with each candidate. It was an opportunity to soak up the conversation between the two of them and learn the significance of various kinds of leaf discoloration, trellising, irrigation, and pruning. Since each conversation covered essentially the same topics, I heard it repeated three times. The repetition was helpful for my learning.

When he completed the interviews, Frank invited me over to help him think through his decision. We sat at the table on his deck, sheltered from the blistering early September sun by a wide, tan umbrella. I filled two glasses from a pitcher of iced tea as Frank slumped in a chair with his hands folded behind his head, looking toward his vineyard through sunglasses. Printouts that detailed his farming costs and grape revenue for the past five years were spread in front of him. All the information he needed was there. Now it was time for him to decide. That was going to be difficult, because none of the candidates had been overly impressive. All were experienced and competent, yet none stood out above the others.

I had never met Frank's current manager, and except for hearing some of Frank's dissatisfied mutterings about a few problems, I wasn't sure why he wanted to change. I sipped my drink and asked how long they had worked together. He cocked his head and paused for a moment. "About a dozen years," he replied.

That was a surprise. "A dozen years? All you've talked about are a few disappointments. They may be annoying, but they don't seem that big to me. When you mentioned them, he fixed them. There's got to be more to the story for you to want to bust up a twelve-year relationship."

"Not really," Frank explained. "Sometimes I can't help myself. He gets under my skin. He's a minimalist. He does what's necessary and that's it. I get nothing fancy." He paused to think a moment before adding, "And sometimes he forgets things, and that bothers me."

"Does he charge you for work you don't think he's doing?" I asked.

"Oh, no," Frank responded. "He's as honest a person as you'll ever meet. It's just …"

He went silent, groping for the right words.

"It's just what?" I finally asked.

"It's just … me, I guess," Frank said thoughtfully. "Every couple of years I get it in my head that it's time to get another manager, but I never find one. Then we work things out and we get along just fine. He's hardworking. And honest, like I said. Knows the business. Has a good crew." His voice drifted off.

"Frank, sometimes change just for the sake of change is … well … not very smart."

"Yeah. I know. But it's always good to look around. Every one of those guys who walked through my vineyard told me something I didn't know. I learned from each one. It wasn't a waste of time. But none of them seemed to know as much as my guy."

"Sounds like you've made your decision."

"I guess I have," Frank said, laughing. "He's my guy … again."

"The next time he comes by, give me a call," I requested. "I'd like to meet someone who would put up with you for twelve years."

Frank threw his head back with a big guffaw. "Sometimes I get unreasonable and can't help myself."

Actually, Frank didn't seem unreasonable at all. He was a good neighbor, a helpful mentor, and an all-around friend. What made it unusual was that we had little in common except our property line. He was younger than I by about twelve years. He was born and bred on the West Coast and I on the East Coast. His father had discouraged him from going to college after high school and mine had encouraged me. He had a dark complexion and a full head of black hair, and I was fair skinned with hair that was gray and gradually disappearing. He had the build of a wrestler and I had one of a basketball player. He was expressive and I was reserved. He was an entrepreneur and I was a product of big corporations. We were a Napa Valley version of the Odd Couple thrown together by circumstances. The experience was working out for both of us.

Chapter 21:
What to Plant?

Laurie Wood had tipped me off to the problem with the Chenin Blanc on his walk through the vineyard in July. Our contract with Beringer specified a minimum Brix reading of 20 and an average of 21.5, Brix being the percentage of sugar in the grape juice as measured by a refractometer. We sampled regularly, but our juice seemed to get stuck on 19.6 in late September and refused to go higher. In mid-October, Beringer took its own sample, which confirmed that our grapes hadn't achieved the sugar level that met their standards. In a follow-up telephone call, they confirmed that they did not want our grapes. Apparently phylloxera and the other diseases Laurie had identified had made the vines sicker than we had realized. Our hope for a cash infusion went up in smoke.

There was some good news, though. The Sauvignon Blanc began to thrive once I followed Laurie's advice and stopped irrigating. The leaves had taken on the vital appearance of a healthy plant and vine growth returned. We would not have a harvest from them for at least a year, but it was good to know they had turned the corner. The popularity of Sauvignon Blanc had waned during the 1980s and early 1990s but now was on the upswing. It thrives in the Bordeaux region and the Loire Valley of France and is the second most popular white wine grape grown in California after Chardonnay. The wines made from it come in a variety of styles but have a common thread of juicy acidity. Flavors of lemon, lime, and grapefruit are very prevalent, but the wine can also have hints of pear, peach, and tropical fruit. I decided that finding a buyer for our crop would be one of next year's chores.

The big decision facing me now was to choose the grape variety that would replace the Chenin Blanc. Historically, vineyard owners planted varieties that they liked or thought they could sell easily or at the highest price. Now, however, research at places like UC Davis and the UC Field Station in Oakville, together with word-of-mouth experience shared among people in the industry, have refined the decision-making process. Climate (particularly average temperature) and soil considerations have become the benchmarks for planting decisions. There is a growing body of information that can be used to help select the most appropriate variety for anyone wanting to use it.

It's difficult to imagine, however, that a new vineyard owner in the Napa Valley would not want to grow Cabernet Sauvignon, particularly in Oakville. It's the King of the Grapes and the primary source of the valley's reputation. Wines bearing the labels of Dalle Valle, Silver Oak, Groth Reserve, Harlan Estate, Screaming Eagle, Opus One, and Heitz's Martha's Vineyard are just a few of the Oakville Cabs that are fawned over and collected like fine art. You drink Zinfandel, Merlot, Sauvignon Blanc, and Chardonnay, but you sip, savor, and "ooh" and "aah" over Cabernet. Was it possible that we had purchased an undiscovered premium Cabernet site that would make us the envy of the wine world? The thought had never seriously crossed my mind, but maybe ... just maybe ... we had a site where it would grow well.

It was possible, of course, to plant Cabernet without first determining if it was well suited for our soil and location. However, habits of a lifetime are difficult to kick, and I had spent almost four decades in business, making decisions based on data rather than hopes or wishes. The only way to be comfortable was to do the research and get the facts. It was too costly a decision to make casually. The facts could always be ignored, but not knowing them seemed irresponsible. Once planted, it is three years before the vineyard has its first harvest and seven to eight years before the vines are at maximum production. Since a new vineyard would have a life of twenty to twenty-five years, it was a decision that had to be fact-based rather than dream-based.

As I had discovered time after time, the valley is rich not only in its wine-growing tradition but in its wine-growing resources. I called Jim Paras for advice. Jim was an attorney in San Francisco who founded the Jade Mountain Winery and had just completed a showplace house on his property near the top of Mount Veeder. We had met years earlier when he did legal work for Marriott in San Francisco. He recommended a couple of names, including Ed Weber at the UC Davis Extension office in Napa, who had already helped me. Ed was his usual accommodating self when I called, and he showed up at my front door two days later.

"You can approach your decision a couple of ways," he explained. "One is just to ask around and find out what varieties are being grown near you. Typically, Cabernet Sauvignon, Merlot, and Sauvignon Blanc do well here. Then there are consultants who know this area pretty well and can advise you. They may suggest you hire someone with a backhoe to come in and do a soil sample, but that starts getting expensive and you are only talking about four acres."

"Then what?" I asked.

"Once you select the variety, you have to select the rootstock. Remember that wine grapes are grafted plants. Rootstock is the part that is below ground, and the variety—it's called a scion—is the part above ground. Different rootstocks have different properties. Some are drought resistant. They are best for places that get little water. Others are best for rich, well-watered soils like you have on the valley floor in Oakville.

They counteract the vigor of the soil and prevent the vine from getting too bushy. Others are disease and pest resistant. Some offer a combination of features. You have to select a rootstock that is compatible with both your site and the variety you want to grow."

Ed was totally at home in a vineyard, but his beard and glasses gave him the appearance of someone you would expect to see puttering around a physics research lab. A native of Southern California, he had gotten a bachelor's degree in Plant Science and a master's degree in Viticulture from UC Davis. An internship at a family vineyard in Tuscany, followed by a job at the Phelps Winery, had prepared him for his position as Napa County's viticulture guru. He was one of those public servants who gave public service a great name. I was a new-comer to the valley with just seven acres of grapes, yet he gave me the courtesy of his knowledge and experience whenever I called, as though I was a big-time operator.

"I've seen D'Ambrosio's vineyard and know he gets seven to eight tons of Merlot per acre," Ed explained. "The vineyards east and north of us are planted in Sauvignon Blanc and Mer-lot. I'm not sure what's planted south of you. I'm guessing it's Cabernet and Merlot. There's been a lot of Merlot planted re-cently," Ed continued. "It's popular, but that could mean supply will begin to outstrip demand soon."

He walked over to his truck, reached into the cab, and handed me a booklet. "This is the Napa County Crop Report that gets published every year," he explained, thumbing through the pages. "It shows the number of acres that have been planted in each variety, including both the bearing and the nonbearing acres. Nonbearing means they are not yet producing fruit but will. Based on these numbers, the amount of Merlot grown in Napa will increase 30 percent over the next three years if no Merlot acreage is taken out of production. Keep that in mind. And remember, there are other issues to think about, too. Like how much space to leave between the vines and how wide to make the rows. I'm sure your vineyard manager is up to speed on them and has his thoughts on what's best."

It seemed like overkill to go through a soil analysis using a backhoe and have the samples analyzed by a laboratory. We weren't planting on a scale that required that expense, but perhaps a consultant could help. I heard that Rollin Wilkinson was a possible resource, so I gave him a call.

He came by that afternoon and we walked through the vineyard. Rollin was familiar with the area because of his job at the Christian Brothers Winery years earlier. He stretched his hand out across a large part of the horizon and said, "We farmed a lot of this land and grew Cab over there," pointing to the northwest. "We got high yields. The quality was disappointing but that's because it was overcropped with more clusters than the canopy could ripen. If you control the vigor and don't try to get too many tons per acre, the quality should be good."

Back at the house, he laid out on the kitchen table a soil map of the valley. It had originally been prepared in the 1930s and had been updated from time to time.

"Over the centuries, sediment washed down from the hillsides and was deposited on the valley floor." Pointing to where we were standing, he added, "Your soil type is Pleasanton loam. It's fertile and well drained. This is a good spot."

"Do you have a recommendation about the variety I should plant?"

"Everyone wants to plant Cab in Oakville and you can, too," he answered, "but I think Merlot will do better. The soils and the amount of heat you'll get during the growing season will produce first rate Merlot."

Operating on the principle that the more you listen, the more you learn, I also contacted Mike Moone, a golfing buddy of a friend from my Marriott days. Mike had resurrected the quality and reputation of Beringer in the 1990s as its president and was part of Silverado Premium Partners, which bought the winery in 1996 and later sold it to Fosters. I reached him by phone and told him my story.

"I'm familiar with where you are," he said. "You made a good buy. Cab should do well but I think Merlot would do better. There'll always be a strong market for the kind of quality you'll be able to grow."

Another person I thought could help was Wayne Hogue, whom I had met during one of my visits to the valley years earlier. He owned, operated, and was winemaker of the Terraces Winery in Rutherford and produced top-drawer Zinfandel and Cabernet Sauvignon. Visiting him in his office, I explained the decision I was facing and my desire to plant Cabernet. As I finished, he leaned forward in his chair and remarked thoughtfully, "I've been in some kind of farming almost all my life. One of the lessons I have learned is that you usually don't go wrong if you take what nature gives you."

I knew what he meant. Merlot was best suited for our soil and location, and Merlot would be my best bet. That did it. Merlot it would be. We'd plant it in the spring, almost ten months away. But I also decided to plant three rows of Cab to see how well it did. After all, conventional wisdom is always conventional but it isn't always right.

Chapter 22:
The Benefit of Walnut Trees

The walnut trees play second fiddle to the vineyard in terms of importance, but they have their own appeal. They offer a degree of agricultural diversity to our property and provide an enjoyable source of shade, beginning early in the morning and continuing throughout the day. The walnuts on them grow in a husk the size and color of a large lime. When they are ripe, the husk splits open like an egg hatching a chick, and the nut in its shell drops to the ground. We called the Diamond Walnut people in the hope of selling our crop, but found no interest. Our orchard was too small to capture the interest of a major commercial business.

Not all thirty-eight walnut trees are Hartleys. Six are un-grafted, and we learned that they are Persians. They don't bear nuts, and assuming they were planted at the same time as the Hartleys, they are aging more gracefully. While the leaves of the Hartleys begin to wilt and look tired in the heat of mid-August, those on the Persians don't lose their crispness. The walnuts begin to ripen and drop in late September, and by that time, leaf drop also starts.

Walnut-harvest times produce an increase in the squirrel population and the frequency of their visits. I'm no sharpshoot-er with my pellet gun, but the squirrels have learned that they venture close to the house at their own peril.

During that first year of living in our new house, Patricia and I discovered how enjoyable it is to harvest walnuts. I say "harvest" but really mean "pick up." In fact, walnut harvesting may have a place in the future of American fitness. First, the physical benefits. Since walnuts blend in with the leaves that gradually accumulate on the grayish-brown soil in the orchard, your visual acuity is sharpened by locating them against this background. Squinting is discouraged. Once located, you pick them up by bending at the waist and feeling your abs tighten and your hamstrings stretch. Done repeatedly, I was sure it could erase inches from my middle and make my legs ready for long-distance running, cross-country skiing, or maybe just walking long distances in shopping malls. Regrettably, these benefits are lost on anyone who uses the long-handled pistol-grip device seen in the hands of highway crews when they remove litter from the side of the road. This motion, however, provides exercise for the hand and forearm while also sharpening the hand-eye coordination required to pluck the walnut from the ground and drop it into a sack.

But it's the psychological benefits that make walnut harvesting unique. It is life at its simplest. To say there is no stress involved is an understatement. Not only is there no stress, but there is also no mental challenge and nothing to interrupt you, provided you have the good judgment to leave your cell phone on the table in the kitchen. Being free of the need to communicate allows your mind to wander. In the quiet and shade of the orchard, I remembered remarks made by others that left me tongue-tied and thought of all the clever rejoinders I wish had come to mind. And I thought of investments I should have made and stocks I should have sold sooner or later than I did. Old pals and former teachers came to mind, as did girlfriends from the past and some of the embarrassingly dumb things I said and did at earlier times of my life. I had the time to plot a strategy for the remainder of my years, but usually spent it thinking about the remaining clues from an unfinished crossword puzzle or what I'd like to have for dinner.

There is a surprising amount of competition for the walnuts. They are, after all, prized by humans, squirrels, and birds alike. It was amazing to watch birds with tungsten-strength beaks hold the walnuts between their feet and peck persistently at the tough outer shell. Eventually they drill through and rescue the walnut meat inside. But all that work! It's tough to know whether the reward is worth the effort, but I don't know what other goodies are on a bird's menu. Walnuts may be the equivalent of filet mignon.

The other competition comes from the people who work in our vineyard and those from neighboring vineyards who use the trees for shade at lunchtime. Property lines have no significance to them, and they rest where it's most comfortable. Shade is rare and much sought-after by vineyard workers in July and August. Walnuts add variety to their normal lunch, and tossing them at a dozing friend livens up lunchtime.

We soon realized that what Laurie Wood had said on his first visit was correct—walnut trees suffer from a lot of ailments. One of the most prevalent is a husk fly that bores into the lime-green husk, turning it black and gooey. Usually the damage done to the husk carries through to the shell of the walnut and makes it unpleasant looking. Sometimes the nut inside is affected and sometimes it isn't, but the discoloration on the walnut shell makes it unusable for all practical purposes. The birds, however, are less selective than humans.

Our husk-fly problem seemed to worsen after we plugged up a hole under the roof of the front deck that led to a nesting place for bats. The bats must have been eating the husk flies. It was a lesson in the balance of nature.

A hazard Laurie had not mentioned was tractor blight. I learned about it on my own. It's a tongue-in-cheek reference to the damage done by careless tractor drivers who drag a disc harrow too close to a tree and skin off bark at ground level. This loss reduces both the tree's protection against insects and the amount of nutrients being transported from the roots to the limbs. The wound heals, but the damage is permanent and shortens the tree's life. Another variety of tractor blight afflicts vineyards, where it takes a toll on end posts, irrigation systems, and the vines themselves.

As the growing season drew to a close in November, we looked back at the progress we had made since May. The house had been painted and remodeled. The walnut trees had been pruned. Before the start of the rainy season, the heavy equipment needed to pluck our Chenin Blanc vines from the ground had done its job and left big piles of vines, trellis wire, and metal posts on our back acreage. The vines had been burned and the other junk had been removed by truck to a re-processor. The earth had been disced as part of the preparation for the spring planting of Merlot. The rains arrived on schedule in late November as the curtain came down on our first year of vineyard ownership. The downs we had experienced were more than offset by the ups and the excitement, anticipation, and promise of next year. It was difficult to imagine a retirement with better prospects.

Chapter 23:
Making Progress

Winter is downtime for a vineyard but a good time for making decisions. I had grown increasingly uneasy with my vineyard manager and wondered whether we were right for each other. For one thing, he had used almost two hundred labor hours farming the Chenin Blanc and had never brought to my attention the fact that the vines were struggling to ripen their fruit. Laurie had done that. I would have expected someone with my manager's experience to notice the problem and point out that we might be wasting money.

Secondly, I learned that he was rapidly becoming a vineyard manager to the "stars"—"stars" being the high-profile and high-net-worth people moving into the valley because of its cache, who had more financial resources than I did. I can appreciate that it's fun to deal with people who appear on the cover of newsstand magazines, but I still expected to get my fair share of time and attention. The widening gap between us was driven home in October when I discovered I had to work through his assistant to schedule a telephone appointment to speak with him.

In anticipation of ending our relationship, I spoke with Frank D'Ambrosio's vineyard manager and learned he would be glad to add me as a client if I wanted to make a change. He had grown up in the valley, and farming was in his blood. The time I spent with him was reassuring. I had already scheduled a meeting with my manager to sever our ties when I received a letter from him explaining that he had a new assistant who would be assigned to work with me. It was someone with extensive viticulture experience with table grapes. That meant I, who knew basically nothing about farming wine grapes, would be relying on someone who had no experience farming them. It was a recipe for disaster, but had the redeeming feature of providing the graceful and polite reason for me to terminate our relationship. He understood, and deep down in my heart I believe he was as relieved as I was.

The vineyard tour with Laurie and the opportunity to sit in on Frank's discussion with prospective vineyard managers had moved me up the vineyard learning curve, but more was needed. Fortunately, UC Davis offers a vast array of courses for audiences that run the gamut from neophytes like me to pros with extensive vineyard and winemaking experience. Looking at the catalog and reading the course description, a three-day viticulture short course entitled "Varietal Grape Production" fit the bill. I attended with my son Brian.

It was conducted by members of the UC Davis faculty and people who, like Ed Weber, were farm advisors in the major grape-growing counties of California. The topics included the theory and practice of canopy management, vine grafting and budding, irrigation systems use, and on and on. It was the proverbial "drinking from a fire hose" experience for both of us. More than anything, it underlined the scale and complexity of our new life and reconfirmed the truth that every business looks simple until you get involved in it.

When the rainy season ended and drilling equipment could once again go into the fields without getting mired in the mud, the Pulliams came back to drill a well at a new site Laurie had selected. I waited with my fingers crossed while they worked. What made the situation so stressful was that I had no good Plan B. If we couldn't find a good water source, we were left with having to grow something other than grapes or building a storage tank and trucking in water. Neither was attractive.

Laurie said we'd hit water at about forty feet and should stop drilling. The Pulliams hit water at that depth and stopped. They estimated the flow to be 32 gallons a minute, which was plenty for irrigating. I took the sample to the lab for testing and was told the water was completely suitable for grapes. I felt the albatross drop from my neck. The Ehrlich Vineyard was in business! Laurie had come through as promised. We brought in a backhoe to make the trenches for the waterlines, installed the new pumps, and our irrigation system was soon up and running.

Since the Sauvignon Blanc would be producing its first harvestable crop in the fall, I wanted to line up a buyer well in advance and relied once again on the wine section of Vallergas Market to identify potential customers. I found eight that looked promising, but I waited for the vines to leaf out in April before I contacted them. Six were interested and seemed anxious to follow up.

Remembering the old chestnut that you only get one chance to make a first impression, I wanted to be well prepared for these meetings. I had acquired enough knowledge to be able to "talk the talk" to a limited extent and avoid appearing to be a complete novice. Fortunately, the Agricultural Crop Report published each year by Napa County contains not only the number of acres planted in each grape variety but the average price per ton paid for them. From it I learned that the average price paid for Sauvignon Blanc in the Napa Valley the prior year was $1,250.

My first meeting was with two young men representing a winery in Rutherford. One was the assistant winemaker and the other's role was never specified. As we walked through the vineyard, they inquired about the empty four acres where the Chenin Blanc had been. I explained we had pulled out Chenin Blanc and would be planting Merlot.

"That's a good decision," the assistant winemaker said.

"What rootstock will you use?" his colleague asked.

"101-14," I answered. My answer reflected a decision my new vineyard manager and I had made during the winter.

"Will it be too vigorous?"

"I don't think so. It's been used extensively in these soils. My neighbor has it and is very pleased with the results. And it has good resistance to phylloxera and nematodes."

He grunted and I waited for a follow-up question. None came, and I breathed a sigh of relief because I had just used up all my rootstock knowledge. As we talked, I mentioned that the vineyard was a new venture for me, and both reacted visibly to that information. It was like a "tell" in a poker game but I didn't know what it meant. After a few more minutes of discussion, they asked to caucus and walked twenty feet away to talk. Their conversation was short and animated. Walking back towards me, the assistant winemaker asked, "How long did you say you've been in the vineyard business?"

"Less than a year," I answered.

"How did you do selling the Chenin Blanc?" he asked.

"I didn't. The vines were too sick to ripen the fruit."

"Sorry to hear that," he responded. "We're interested in your Sauvignon Blanc. This is a good spot. We're looking at a couple of other vineyards this afternoon but are willing to offer you a contract at $950 a ton. That's a good price."

A good price? I knew that was nonsense. Now I understood their earlier reaction. They were offering me a price that was more appropriate for Thompson Seedless grapes. Apparently they assumed that since I was new, I must be uninformed and could be taken advantage of. They totally missed the fact that as a businessman I knew the basics of negotiating, the first rule of which is to know the value of what you're selling.

"That's an interesting offer," I said noncommittally. "I'll have to think about it."

"Do that if you want, but we're visiting a couple of other growers this afternoon and … well … once we get enough fruit under contract, our offer is off the table."

"I guess I'll just have to take that chance," I responded.

They hesitated. "It's a good price," one of them repeated.

"I wouldn't expect anything else from a winery with your reputation. I'll let you know."

We shook hands and they drove off.

It was not a good beginning. This is going to be grueling, I thought. What annoyed me was their assumption that being new to growing grapes meant I didn't understand how business was conducted. I brushed off the experience as just another part of the vineyard-initiation process, like writing lots of checks.

That afternoon I met with Michael Weis, the winemaker for the Groth Winery, located east of us on the Oakville Cross Road. Michael had a scholarly appearance packed onto an athletic frame. He arrived with the owner, Dennis Groth. Dennis had been one of the architects of the wildly successful Atari Corporation and had purchased 121 acres of land in Oakville in 1981. After Atari was sold three years later, he and his wife Judy opened their winery and became Napa Valley residents. He looked like he would be equally comfortable in a board room or a vineyard.

"You've got a nice spot here," Dennis volunteered pleasantly. "It's nice to meet the neighbors."

"We enjoy being in Oakville."

"What made you decide to buy in the Napa Valley?" he asked.

I gave him a short, uncomplicated version that emphasized our affection for the place. I also included an explanation of our empty acreage and the plans for planting Merlot.

"We'd like to look around, if you don't mind," Michael said. "Just wander. Get a feel for the soil and your trellis system."

"Please do. I'll be glad to answer any questions or get answers from my vineyard manager."

They walked between two rows, looking at the trellis system and the soil. They actually kicked the dirt with the toes of their work shoes in a way that reminded me of the "kicking the tires" expression used for car buyers. At the end of the row, they split up and walked back between different rows and then conferred just off the back deck. I joined them.

"Your vineyard looks like it's in the hands of a good farmer," Michael remarked. "We'd like to offer you a contract for $1,350 a ton. That's $100 above last year's Napa average. Our contracts are short and uncomplicated. They have a formula for year-to-year price adjustments based on the Napa average and provide for two-year notice of cancellation by either party after the first year."

The difference between their approach and the one taken by my earlier visitors could not have been more pronounced. It was straightforward and no nonsense. They knew I was new to the business, yet they were willing to pay a fair price and provide a fair contract, not just mouth those words.

"What you say sounds attractive," I responded, "but I'm new at this game. I'd like to run it by my vineyard manager, since he's much more familiar with this than I am."

"Not a problem," Michael said. "Let us know in a week or so. If you're not interested, we'd like to know as soon as possible."

"Absolutely."

When they left, I went inside to call my manager and give him an update on the two meetings.

"The Groth offer is solid," he said. "That's a good price.

I know the kind of contract they're talking about. It's pretty standard. Just be sure to read it over to make sure it says what they said it says."

"I'm lined up to speak to three other wineries, but it's difficult to think I'll get anything much better than that. Plus, I really like these guys. And they're in Oakville."

"That's your decision."

My next call was to Michael to confirm my interest and ask to see the contract. He delivered it the next morning, and it was just as he had described it. There was no reason to wait. I signed and returned it to him. We shook hands, and I went home to cancel my meetings with the other prospective buyers. The income was months away, but now we had a buyer for our fruit. It provided a financial backstop of sorts and put us on the road to becoming a business rather than a money pit.

Chapter 24:
Getting Established

In May, we planted 101-14 rootstock in the new vineyard. It had become one of the workhorses of the valley because of its suitability for most soils and its resistance to phylloxera. This vineyard was thirty-eight rows wide and ninety-one vines deep. After it had grown for a year, we would bud a Merlot scion onto thirty-five rows and a Cabernet Sauvignon scion onto three rows and, two years after that, we'd get our first red wine grape harvest.

The growing season was a bit cooler than normal but the Sauvignon Blanc vines thrived. The soil retained sufficient water from the winter rains to sustain growth without irrigation until mid-July. The response of the vines to irrigation made it clear the boron problem was a thing of the past. My takeaway from our first growing season is that nature can't be rushed and has its own beat and rhythm. Our first harvest on September 11 produced four and a half tons. It felt good to be there when the delivery was made to Groth and even better when the check for payment arrived.

When you sell grapes, the person you have to satisfy is the winemaker who buys them. Fortunately, Michael Weis liked our fruit from the start. As we got to know him, we learned that his original career choice was not the wine industry. Michael was a graduate student at UC Davis in 1971 headed toward a career in microbiology when he experienced the pleasure of drinking a glass of Robert Mondavi's Fume Blanc, the name Robert had given to his Sauvignon Blanc. Mondavi made the name change in the late 1960s to give a flair and distinction to a wine that had fallen into disrepute among consumers, revitalizing its reputation. Michael's epiphany, as he describes it, caused him to redirect his career to winemaking. His first job in the wine industry was as an experimental enologist at Mondavi and he joined Groth in 1994. My first impression that Dennis and Michael would be good to work with was correct. They are a first-class team in every respect. Dennis, his wife Judy, their children Suzanne and Andrew, and Michael Weis have a strong commitment to quality that is evident in the outstanding wines that bear the Groth label.

Owning a vineyard in Oakville and selling our fruit to an Oakville winery gave us a foothold in the Oakville wine community. We had already met the Swansons and the D'Ambrosios and had known Nils Venge from years earlier, so we had the start of a network. Advice on how to expand it was provided by Michael, who suggested we join the Oakville Winegrowers Association. Formed in 1993, its role is to preserve, promote, and protect the Oakville appellation that was approved that same year. Appellations are wine-grape-growing regions in the United States designated by the Tax and Trade Bureau (TTB) of the Treasury Department. It is a system modeled after the Appellation d'Origine Controlee laws of France.

Membership in the Association, abbreviated as OWG, includes both the big names and the small names in Oakville winegrowing, all of whom have a strong commitment and dedication to their work. They are friendly, generous with their time, and willing to share their experience and knowledge. The big event on the OWG calendar each year is the dinner, which the major wineries take turns hosting. Each attendee brings a bottle or two of his or her best wine, to avoid, perish the thought, the possibility that someone should have an empty wine glass during the evening. While the dinners are always enjoyable, the most memorable one in recent years was held on the patio of the Groth Winery in September 2001 to honor Robert Mondavi. None of us knew it at the time, but the world would be shaken three days later by an event known simply as 9/11.

Casual dress is typical of these events, but this one was formal, befitting the occasion. Patricia and I attended with our daughter Susan. The invitation from co-chairs Dennis Groth and Clarke Swanson explained that each member should come prepared to say a few words about the guest of honor. Clarke was the master of ceremonies, and each speaker he called to the podium described how Robert had provided assistance, advice, or encouragement to him or her and how much Robert had set the tone for cooperation and the standard for quality in the Napa Valley.

My reaction after receiving the invitation was that I would attend but not participate. While I had met Robert on a few occasions, it would have been a stretch to say I knew him. However, as the event grew closer, it struck me that I knew him by reputation and that was good enough. I put my thoughts into a poem that followed the meter and rhyme scheme of Joyce Kilmer's "Trees." When it was my turn, I mentioned the challenge of talking about someone I barely knew and who was known so well by others and then read my poem, explaining that it was tailored after Kilmer's masterpiece:

No poem could ever be as fine

As a Bob Mondavi wine.

A Cab, a Zin, a Chardonnay

Each one superb in its own way.

All are a tribute to the man

And others of his famous clan.

Some said he was too starry-eyed

With views that they could not abide.

With gusto he did prove them wrong

Till now all sing his siren song.

Fine wine's now made by quite a mob

But only God can make a Bob.

As I was walking back to my seat, Jean Phillips, the founder and owner of Screaming Eagle, put her hand on my arm as I passed by and whispered, "You may be a newcomer, but you were on the mark."

The following Monday I got a call from Robert's secretary, asking if I would give him a copy of the poem. I was flattered and explained it would take me a few days to get a copy printed and framed. We had two house guests, Dave and Lorry Rolston, the day he could see us and they came along. Robert appreciated the nicely framed gift I presented to him and, in his gracious manner, sat and discussed the wine business with us like an old friend. He talked candidly about the challenge of trying to decide which of his sons would succeed him when he retired. As it turned out, he was never able to come to grips with that issue and his company was eventually acquired by Constellation Brands in December 2004.

Two years after the OWG dinner, the poem was accorded an unexpected prominence during the Mondavi Winery celebration of Robert's 90th birthday. As part of the occasion, the walls of the Vineyard Room were covered with citations, awards, and miscellaneous memorabilia from his life. The skylights in the twenty-foot-high ceilings make it bright and cheerful. Three of its walls are stucco and the other is glass and looks out on the vineyard. Sweeping arches on three of the walls add grace and dignity to the room, but it was on the archless south wall that the most prestigious awards were placed. In the center was a large black frame containing an award from the Chevaliers du Tastevin. Written in an elegant flowing script, it projected the importance of this world-famous group from Burgundy. Next to it on the left was a colorful State of California citation that honored Robert and extolled his contributions to the California wine industry. Next to it on the right, neatly typed and framed, was the poem I had given Robert. It was in select company.

Chapter 25:
Cabernet Gets the Nod

We grafted the Merlot and Cabernet scions to the 101-14 rootstock in May 2000 and two years later were ready for our first harvest. Rutherford Hill Winery had agreed to buy the Merlot, but the Cabernet was a different story because the amount we produced from 273 vines was too small to sell. What to do? The solution came at the annual dinner of the Oakville Winegrowers Association in 2002. Since Robert Mondavi had agreed to host it at his home, it attracted eighty people, twice as many as usual. He and his wife Margrit Biever greeted everyone like old friends, welcoming us with warm hospitality.

At dinner, I mentioned to those at our table that we were preparing for our first Cabernet harvest. It was going to be less than half a ton, and I wondered if anyone knew how we could have it made into wine for our own use. Nils Venge of Saddleback Cellars, the winemaker of that groundbreaking 1985 Groth Cabernet who was sitting across from me, said it sounded like we might have enough for 25 cases. He casually volunteered to do it. No big deal. I almost fell off my chair. We made a handshake deal before the evening ended, and a few weeks later Patricia and I helped his crew harvest our crop. They transported it less than a mile to his winery where it was crushed, fermented, and aged in a newly purchased French oak barrel.

Nils's assistant, Jeff Fontanella, produced an excellent wine that our family gathered to bottle, cork, foil, and label two years later in December 2004. Jeff developed his winemaking skills as an intern at Opus One while attending UC Davis and, after graduation, worked at ZD Winery before joining Saddle-back. It was a treat to be able to put an Ehrlich Vineyard label on those bottles and make them available to family and friends. We took everyone on the bottling crew out to lunch to cel-ebrate and gave each of them a bottle or two for their efforts. Nils allowed us to follow the same routine in subsequent years and it became apparent to Nils, Jeff, and the Ehrlichs that the Cabernet we grew produced good wine. Based on that, we de-cided to pull out half the Merlot (two acres) in the fall of 2005 and replant with Cabernet in the spring of 2006. By this time, Jeff had opened his own winery on Mount Veeder in Napa and was using his winemaking skills and experience to build a thriving business.

Our good experience with the Cabernet Sauvignon scion we used in our three experimental rows was proof enough that it should be used in this replanting. It had lived up to its repu-tation for producing an aromatic, full-bodied, well-balanced wine with deep color and rich flavor. But what rootstock and trellis system would be best? Both are keys to producing top quality grapes.

The importance of rootstock is obvious: it links the scion to the soil. The one you select has to take into account the fertil-ity, chemistry, and water retention characteristics of the soil and the need to resist pests and disease. There is a dizzying number of choices that are available, each with a unique set of qualities that distinguish it from the others. In deciding which to choose, we had the good fortune to have the guidance of Andy Walker at UC Davis who is one of the world's experts on rootstocks for winegrowing.

I had changed vineyard managers a few years earlier and Andy was acquainted with Paul Garvey, my manager, and his brother Pat who also is prominent in the vineyard management world. Andy volunteered to help us by visiting our property, taking soil samples, and sharing his advice and wisdom. His counsel was realistic and practical and it seemed clear that the selection he recommended would give us the best chance of growing excellent fruit.

The trellis system decision was easier. We were already using a quadrilateral system in which shoots get trained onto parallel wires on each side of the trunk. The shoots mature into cordons on which the cluster-producing buds are grown. The canopy it produced had been effective in intercepting and absorbing sunlight to ripen the clusters and also providing the shade that prevented them from getting sun-burned. During Ed Weber's seminar on trellising that had been part of the UC Davis program, he said there is no "best" system. Our quadrilateral system was old school, but the posts and wires were already in place and it had served us well. Why change? We decided not to.

In May 2006 we planted rootstock onto which the scion had been grafted a year earlier by the nursery. That meant we'd have our first harvest in two years rather than three.

Of course, the real test of any scion and rootstock combination isn't what is written about it but how well it does when it's planted in your soil. Our decision was a leap of faith, an educated one of course, but still a leap of faith. We crossed our fingers, hoped for a soft landing, and were encouraged by the energetic vine growth and the profusion of small, delicate leaves we saw as the growing season wore on. We were off to a good start.

Chapter 26:
Keeping the Orchard Trimmed

The importance of the vineyard is always front and center, but the trees on our property also require care and attention. The age of the walnut trees and the combined weight of their branches, leaves, and nuts cause the limbs to droop, almost touching the ground. Droop, as we all discover, is a challenge to living things as they age, and it gave the orchard a sloppy and unkempt appearance. However, with a stepladder, a hand-saw, and a pole saw, I was able to trim the trees and keep them neat. It's enjoyable work. The limbs I removed were dragged to a burn pile located at the western edge of the property that got torched in late autumn on one of the days authorized for burning by local air-quality authorities.

The cutting and dragging of the limbs add to the orchard's contribution to fitness. It provided a degree of aerobic exercise that millions of Americans get by exerting themselves in health clubs. I was ahead of the game because it cost me nothing, except for the periodic replacement of saw blades, and it occurred in the fresh air with no one sweating on the next machine. Truth be told, part of the motivation for removing low-hanging limbs comes from the need to have a disc harrow dragged through the orchard every spring to turn under the weeds and break up the soil to improve drainage. If the limbs hang too low, the tractor can't get under them to do the job.

The big challenge to a tree trimmer is deciding where to stand when you're cutting so you are not the victim of a falling limb, particularly a thick one dropping from a height of ten to fifteen feet. It's a talent you pick up quickly or you decide to spend your time doing other things. My confidence in dropping limbs where I wanted them to drop eventually grew to the point where I felt comfortable expanding my repertoire to include oak-tree trimming. In particular, one of the sixty-foot-tall oak trees near the western edge of our property had a low-hanging limb that needed to come down. To get to it, I had to stand on a high rung of my stepladder and use a robust chainsaw rather than my less sturdy pole saw. Ordinarily, I would hire a professional to do a job like this, but it was only one limb and I convinced myself that taking it down was in my wheelhouse.

I approached the project with the care of a surgeon. I decided how much of the limb I wanted to remove and decided it would be best to do it with two cuts. I decided to position the ladder on the driveway in a place where the falling limb would not be a hazard to me. Confident in my assessment, I fired up the chainsaw, climbed the ladder to the fourth rung, and began cutting. The saw was sharp and cut through the six-inch-diameter limb with ease.

It was a third of the way through the limb when, for some unknown reason, it gave way and swung toward me rather than away from me. Its weight and momentum hit the ladder and knocked me backwards off it. I instinctively took my finger off the saw's trigger, shutting it down. I remember falling and then remember looking up at tree branches. I was lying on my back but, since I didn't remember hitting the hard ground of the driveway, the fall must have knocked me out. It took a couple of seconds to regain my senses and reconstruct what had happened. When I did, I realized I'd be better off to confine my tree trimming to walnut trees. I also realized that one advantage of living where you are remote from your neighbors is that there's little likelihood the dumb things you do will be seen by anyone.

Olive trees have become increasingly popular in the Napa Valley and the dozen we have were planted by the prior owner for their landscape value. They often share the land with vineyards in Tuscany and contribute to those beautiful landscapes associated with the Italian countryside. The fruit on our trees ripens between mid-October and mid-December but has been harvested only once. A neighboring winery with an extensive number of olive trees agreed to take our fruit along with theirs and have it pressed into oil for their employees. We got half a dozen bottles as our share. It was a great swap.

Because the shape of olive trees is reminiscent of weeping willow trees, pruning them is an art form. A few years after we moved in, I needed some pruning advice and thought of the Round Pond Olive Press in Rutherford. The manager of their orchard and the Round Pond vineyard is Chris Pedemonte who, with responsibility for hundreds of olive trees, is as knowledgeable about them as anyone in the valley. I gave Chris a call and what occurred next is typical of the spirit of cooperation that I've experienced time after time with people in the Napa Valley farming community.

He listened to my request for assistance and asked where I was. I explained that I was at my ranch, which is about ten minutes from Round Pond. "If you have the time, I'll come over and take a look," he volunteered. I assured him I'd be here. He showed up twenty minutes later and checked out my trees. "They look like tree balls," he observed. "When people don't really know how to prune them, this is what happens." He explained that they have to be pruned "so they drape and birds can easily fly into them," and described what that meant.

"Do you know anyone I could hire to do that?" I asked.

"Sure," he said. "One of the guys on my crew probably would be glad to make some extra money. I'll check and get back to you."

He departed with a handshake and called an hour later to explain the arrangements he had made with a person on his crew to prune our trees. I had spoken to Chris by phone once three years earlier. He really didn't know who I was, he owed me no favors, and he had nothing to gain by offering assistance. He did it because helpfulness is in his DNA and because the Napa Valley farmers I've met don't hesitate to assist one another. People who gravitate to farming have an attitude and outlook that is different from people in other commercial enterprises. My forty-year business career with two large, successful companies made me acquainted with plenty of aggressive people who sought to control their environment and who strived to shape their destiny. Farmers understand they don't have much control over those things because Mother Nature has the upper hand. Since they are always wrestling with her, they go out of their way to help one another. I have repeatedly found myself the beneficiary of that outlook.

Chapter 27:
The Oakville Story

Oakville didn't top my priority list when I started my real estate search, but it certainly has proved to be an outstanding place to have a vineyard. The modest sign that heralds your arrival there is like all the others along Route 29: white lettering on a green background. Driving up from Napa, the sign appears just before you reach the Oakville Grade, a road that takes you west on a harrowing ride up and over the Mayacamas into Sonoma.

Unlike other parts of the country, where a sign may say "Welcome to the Heart of the South" or "The Friendliest Town on the Platte," the white block lettering on the Oakville sign simply informs you that it has a population of 300 and an elevation of 150 feet, perhaps giving Oakville the distinction of being the only community in America with exactly two residents for each foot of elevation.

However, I began to doubt the accuracy of the population figure after living in Oakville for a few years. I couldn't imagine where those 300 people lived. My quick count left me far short. That's when my imagination got me wondering if Oakville had a "Brigadoon" quality to it.

"Brigadoon" is about a Scottish village that appears every hundred years for one day and then disappears into the Highland mists for another hundred years. (Film buffs may recall that, in the movie version, American tourist Gene Kelly sings and dances to "Almost Like Being in Love" after he meets villager Cyd Charisse.) Did Oakville have a mysterious population explosion every decade when the census was taken, only to shrink to its normal size when it was over? Hmmm....

Fortunately, local historian Lin Weber, writing in the Spring 2013 issue of *Tidings* published by the Napa County Historical Society, set the record straight with the statement, "In 2010 there were 71 Oakvillians, despite a sign that claims 300 residents." Her number was in the general neighborhood of the one I had estimated.

Whatever its population, Oakville has acquired a worldwide reputation for its Bordeaux varietal wines, particularly Cabernet Sauvignon and Cabernet Sauvignon-based blends. For those new to the wine world, Bordeaux is the wine region in southwest France that makes red wine predominantly from Cabernet Sauvignon, Merlot, Cabernet Franc, Malbec, and Petit Verdot grapes and white wine predominantly from Sauvignon Blanc and Semillon grapes.

The history of winemaking in Oakville begins with Hamilton Crabb, a farmer from Ohio who moved to California in January 1853 to try his hand at prospecting in the Sierras. When that didn't work out, he returned to farming in Amador County. At the urging of John Lewelling, a friend who had discovered the beauty of the Napa Valley, he purchased 240 acres in January 1868 "situated at Oakville, on the line of the railroad, twelve miles north of Napa City." He called it Hermosa Vineyard. Graeme Macdonald, whose family has owned land in Oakville for five generations, says Crabb purchased the land from a descendant of George Yount. This was at a time when there were only a few hundred acres of grapes in the Napa Valley and the most dominant use of land was for wheat farms and pastureland.

Crabb opened a winery in 1872 and by 1884 it was the largest one in the upper Napa Valley, having a capacity of 400,000 gallons. He purchased adjacent land over the years and applied the name To Kalon to his holdings, a Greek term meaning "most beautiful." It became the brand name he used to develop a national market for his wine as well as the name he applied to his vineyard, saying, "I try to make it mean the boss vineyard." Using current landmarks, Crabb's original vineyard ran west from Highway 29 toward the Mayacamas and north from Walnut Drive. The additional acreage he acquired in 1881 moved the southern border from Walnut Drive to the Oakville Grade.

Crabb introduced experimental viticulture to the Napa Valley in the mid-1870s when concern about phylloxera caused him to acquire several hundred native vine cuttings from George Husmann, a viticulturist in Missouri. His passion for the subject led to the creation of the largest collection of vinifera grape varietals in the country. His advice and vine cuttings helped fuel the valley's first great viticultural boom during the 1880s.

By 1890, Crabb's prominence reached the point where the *Chicago Herald* called him "the Wine King of the Pacific Slope." He was an early proponent of Cabernet Sauvignon, and the wines he produced were of such high quality that he won eight awards at the Columbia Exposition in 1893. Crabb, of course, was not alone in attempting to establish Oakville as a wine center. Adolph Brun and Jean Chaix built a winery in 1877 on what is now part of the Napa Wine Company site and John Benson began construction of Far Niente in 1885.

Crabb's death on March 2, 1899, was a loss to Napa Valley viticulture. Unable to pay off a loan he incurred just a few months before his death, his family sold his property at public auction. However, his legacy was perpetuated when the Department of Agriculture acquired a portion of his experimental vineyard in 1913 and deeded it to the University of California in 1954 to create the UC Davis Experimental Station.

While the Napa Valley was still recovering from its post-Prohibition funk, Martin Stelling, a wealthy San Franciscan, assembled thousands of acres in Oakvile in 1943, including most of Crabb's original To Kalon property. His commitment to the cultivation of high quality varietal plantings led Andre Tchelistcheff to call him "the new oxygen" of the Napa Valley. However, his untimely death in an automobile accident in 1950 cut short his legacy. His heirs sold off some of the land, but their plans to build clusters of luxury homes in To Kalon ran into strong local opposition and were never approved.

Eventually Robert Mondavi acquired much of the To Kalon vineyard for the new winery he opened in Oakville in 1966. It was a shot in the arm to Oakville and the valley in general. The partnership he later formed with Baron Philippe Rothschild of French wine fame to create Opus One further enhanced Oakville's stature in the wine world.

As the home of well-established Cabernet Sauvignon stars such as Harlan Estates, Screaming Eagle, Opus One, Dalla Valle, Far Niente, Groth Reserve, and Nickel & Nickel, and outstanding up-and-comers like B Cellars and Tierra Rioja, Oakville is a classic example of the American tradition of each generation standing on the shoulders of the preceding one. Knowledge and skill get acquired, passed along, refined, revised, and improved. The foundation laid by Hamilton Crabb and strengthened by Martin Stelling and Robert Mondavi represents a merger of agriculture and science that has benefitted Oakville's current winegrowers and, perhaps more importantly, wine drinkers everywhere.

Chapter 28:
Getting Past a Bump in the Road

Laurie Wood's observation the first day we met that grape-vines are hearty plants certainly rang true as I watched the newly planted Cabernet grow. At the outset, grow tubes protected the vines from animals and wind, and their greenhouse-like effect provided some moisture at the base of the vine. They were removed in August once the vine's upright growth was well-established.

Shoots sprouted from the vine and, because we were using a quadrilateral trellis system, four of the healthiest were tied onto fruiting wires. During dormancy that winter, the other shoots were removed. The four tied to the fruiting wires became cordons on which spurs developed that produced grape clusters in 2008. Our first harvest was about three tons. Half a ton went to Jeff Fontanella to produce Ehrlich Vineyard wine for family and friends, and we sold the remainder to a local winery.

In June, prior to the harvest, we celebrated the tenth anniversary of Ehrlich Vineyard. The one hundred attendees included many of the people who had given us a lift in the early days: Laurie Wood, Tom and Aaron Pulliam, Guy Larsen, Frank D'Ambrosio, Becke Oberschulte, Filiberto Chavez, and Nils Venge. Our good friends, Mike and Marilyn Riehl, traveled from Bethesda, Maryland, and Patricia's brother, Al, and his wife, Thu, drove up from Saratoga to be there. Brian and his wife Christina, Susan, and Scott were joined by friends from across the country. The walnut and oak trees provided

much-needed shade from the blistering sun. We served an assortment of hors d'oeuvres accompanied by Groth Sauvignon
Blanc and Ehrlich Vineyard Cabernet Sauvignon, but the 100+
degree temperature made water the most popular beverage.

During the planning, I made a tongue-in-cheek remark
that we needed a brass band. Once the thought was expressed,
however, it gained traction. We started a search and, before
long, discovered the existence of the St. Helena Community
Band and arranged for them to serenade us. In the middle of
the afternoon, musicians appeared in the driveway in front
of the house, set up their chairs and instruments, and played
"Stars and Stripes Forever" and an assortment of other peppy
tunes. They made it Oakville's most festive occasion of the year.

Our hopes were high the following spring for a crop of four
to six tons, only to get a sucker punch from Mother Nature in
the form of a pest that was making its appearance in the Napa
Valley for the first time. It was discovered in our vineyard by a
winemaker who, as a prospective buyer, was walking through
it with Paul Garvey and me. Its calling card was to turn ten
or twelve grapes on each cluster it infected into mush. In the
shade cast by the canopy, the damage was almost impossible
to detect visually. You had to touch the clusters to feel it. We
didn't know what it was. Back at the house, I went through
a copy of the vineyard owner's bible on pests – appropriately
named *Grape Pest Management* – and followed up with some on-
line research. It looked like something called Orange Tortrix.
Fortunately, it did not affect all clusters but damaged enough
of them to take the wind out of our sails – and sales.

Other vineyards were affected, too. Monica Cooper, the
University of California Farm Advisor for Napa County,
eventually identified it as the European Grapevine Moth and
initiated prevention and control methods that quarantined the
problem and substantially curtailed its damage in the following
years. Since most of our vineyard was not infected, we were

able to sell a carefully selected ton and a half of good fruit to the winemaker who discovered our problem and we sent half a ton to Jeff for Ehrlich Vineyard wine.

It was a relief to discover that the European Grapevine Moth did not make a return appearance in our vineyard in 2010. A good growing season produced six tons that we sold to another local winery, and in December we bottled the Ehrlich Vineyard wine from the 2008 harvest. While I had liked the wine we produced in 2002-2005 from our three experimental rows, it always fell short of my highest expectations. We had been tasting the 2008 as it aged in medium-toast French oak barrels, and it was apparent by late 2009 that the wine produced from this new rootstock-scion combination was a big improvement. When Patricia and I tasted it with winemaker Jeff Fontanella just before bottling, we knew we had a winner. It was mellow, fruit forward, and rich tasting. It filled your mouth and showed real promise for future improvement from ageing. We bottled 23 cases.

While Napa Valley weather is reasonably predictable from year to year, there are still variations that prevent the vintages from being carbon copies of one another. An example occurred in 2011. The cool temperatures that year caused the harvest to be later than usual, with much fruit being harvested at a Brix level that was lower than winemakers thought produced the best wine. In many cases, alcohol levels for Cabernet Sauvignon were in the moderate 13.5-14.5 range rather than pushing 15 as they did in a normal year. The result is wine considered to have the finesse and elegance of a European wine.

When I hear people comparing wines and vintages like so many did after 2011, I always think back to a conversation I had with Wayne Hogue, the owner of Terraces, a decade ago. I had dropped in on him to pick up a case of Cabernet and congratulated him for having received a 90+ rating in a recent edition of *Wine Spectator*. "Thanks," he said with less enthusiasm than I had expected. "Doesn't that please you?" I asked. "Oh, sure," he answered. "It's flattering, but sometimes people put too much emphasis on what 'the experts' say. Taste is personal. What's important is how it tastes to you. I always thought that if I liked a wine, it was a good one. If I didn't, it wasn't." And this from a man who had learned winemaking from Chuck Wagner Sr. at Caymus and who had an impeccable palate. Wine tasting doesn't get any more straightforward than that.

By the start of 2012, the Ehrlichs had the beginnings of a flourishing vineyard business. With our Cabernet production not yet topped out, the backbone was still Sauvignon Blanc. It was a reliable producer of slightly less than 20 tons of high quality fruit a year, a ton of fruit being the amount needed to produce 60 cases of wine. The demand for Merlot had dropped, but we were fortunate to have a buyer for all of it at a fair price.

In the course of following up with wineries interested in our Cabernet, Paul Garvey learned that Kirk Venge was in the market for a new source of Cabernet for B Cellars where he was the consulting winemaker. Paul has a deep respect for Kirk's father, Nils, and knew I did, too. Kirk, a UC Davis graduate and a classmate of Jeff Fontanella, is cut from the same cloth as Nils – likeable, talented, hardworking – so the possibility of working with Kirk and selling to B Cellars moved to the top of our priority list. I wasn't familiar with B Cellars at the time, but learned it had been founded in 2003 and was producing its wine at another winery's facility. Paul and Kirk eventually negotiated a one-year contract for most of our fruit and we had no trouble finding buyers for the balance.

During the year, I became curious about my genealogy and began researching it, using *Ancestry.com*. All I knew was that my mother's parents had emigrated from Poland in the early 1900s and my father's grandparents had emigrated presumably from Germany in the mid-1800s. I say "presumably from Germany" because ehrlich is the German word for honest. I had some initial success on my own but then hit a stone wall and decided to contact Tristan Tolman, a genealogist in Highland, Utah, who had been recommended to me by a friend. Tristan waded into the project with what I learned was her customary gusto, enthusiasm, and skill. Before long, I had a picture of my ancestors that revealed something unexpected.

Great-grandfather Ehrlich had arrived at Ellis Island on November 11, 1871, on the steamship Bremen from Bremen, Germany. The ship's manifest for the steerage passengers included their name, age, country of origin, and occupation. It is from the manifest that I learned that Joseph Ehrlich, age 27, was a gardener. That meant the distance between me and someone in my lineage who worked the land was only two generations. Between him and me were his son (my grandfather, who was a New York City firefighter) and his grandson (my father, who was a New York Telephone Company engineer). Querencia, the sense of being nourished by the place to which you belong, had steered me to the Napa Valley, and something in my genes had led me to become a vineyard owner.

Chapter 29:
Hitting Our Stride

The 2012 harvest for Ehrlich Vineyard started, in typical fashion, with Sauvignon Blanc. The harvest crew arrived at 5 a.m. on September 6 and went right to work cutting the clusters from the vines and dropping them into plastic picking baskets. As each basket filled, it was emptied into the white bins provided by Groth that each held half a ton of fruit. Truckloads of either six or eight bins were then driven over to the winery and weighed. When the deliveries reached twenty-eight tons, Michael Weis good-naturedly asked the driver, "Where's this stuff coming from?" The year before he had gotten eighteen tons from us.

One of the unique sounds of the Napa Valley is experienced only during harvest time. It's the plop made by ripened grape clusters falling into picking baskets. It's a sound to which vineyard owners are particularly attuned. In the quiet of early morning and over the chatter of the crew, a vineyard owner hears it and experiences the relief of knowing that the days of negative cash flow are numbered. After bankrolling the farming of grapes since the last harvest, it's reassurance that payday, and positive cash flow, are close. That morning there were a lot of plops heard in the Ehrlich Vineyard. By the time we were finished, Groth had received slightly more than thirty-two tons.

What no one knew at the time was that the 2012 harvest would be one of the most bountiful on record in Napa County. The production of white wine grapes was 58 percent higher than in 2011, and the production of red wine grapes was 47 percent higher. Combining the two, the harvest was 50 percent bigger. Making it all the more unbelievable was the fact that the producing acreage in 2012 was down slightly from 2011. Yields were through the roof. It was one of those years that vineyard owners dream about. The fourteen tons of Merlot picked on October 14 (up from ten tons in 2011) and the seventeen tons of Cabernet picked between October 27 and 31 (up from eight tons in 2011) combined with the Sauvignon Blanc to give us a record-breaking total of sixty-three tons.

Again, in December, we bottled Ehrlich Cabernet. This time it was the 2010 vintage. The 2009 that we bottled the year before had been a disappointment. The damage done by the European Grapevine Moth was apparently more extensive than we realized and the wine was not very good. We decided to give it more time to mature. Unfortunately, that didn't seem to make a difference so Patricia and I eventually took 21 cases of it to a corner of the walnut orchard and poured the contents of 252 bottles onto the ground. It made no sense to keep wine that neither we nor Jeff were proud of. The 2010, on the other hand, was even better than the 2008.

We were pleased that B Cellars had found our fruit to their liking, and Paul and Kirk reached an agreement early in 2013 for most of our Cabernet. By then, the owners of B Cellars had purchased a horse ranch on almost twelve acres of land on the Oakville Cross Road. After getting their operating permit from Napa County, they broke ground for the construction of their new winery in June 2013. We were going to be neighbors.

Nature's kindness to vineyard owners continued in 2013 with county-wide tonnage slipping only 4.4 percent from 2012's bonanza. Groth received another through-the-roof harvest of SB from us while our Merlot and Cabernet yields were down slightly. Within a year, Robert Parker unexpectedly put his imprimatur on the 2013 harvest, saying that it produced the finest wines from Napa and Sonoma that he has ever tasted in 37 years.

B Cellars had two years of experience with our fruit by the start of 2014, and Kirk began discussions with Paul about a long-term contract that was hammered out relatively quickly.

Shortly thereafter, I decided I wanted to meet Duffy Keys and Jim Borsack, the owners of B Cellars. It was clear from the start that they came from a different mold than most of the winery owners I knew. I learned that their successful business careers had given them extensive experience in selling and marketing luxury items. They had met at a barbeque in 2002 at a time when each was trying to decide how to get into the wine industry and decided to forge a partnership.

They believed that the future of the wine business would be at the premium end of the scale, which, in turn, made it brand and marketing driven. Think Bryant, Screaming Eagle, and Harlan Estate. They wanted their wines, their winery, and the experience of their guests to reflect upscale quality and luxury. The template they were using was entirely consistent with Duffy's experience as an executive with The Four Seasons hotel company and Jim's experience as an owner of El Portal, a chain of designer leather goods. By taking an approach to the winery business that featured not only high quality wines but high quality food and wine pairings in an upscale setting, they seemed to be as close to the cutting edge of change in 2014 as Robert Mondavi had been in 1966 when he opened a winery that was customer-friendly, emphasizing a totally new and exciting consumer experience.

In the middle of the 2014 growing season, the lower portion of the Napa Valley was rocked by a 6.0 magnitude earthquake in the early hours of Sunday, August 24. It damaged a number of older buildings in Napa, many of which, including the post office on Second Street, were later condemned. One of the wineries most affected was the Fontanella Winery on Mt. Veeder whose barrel room was tumbled and tossed like no other. Jeff, who is always resourceful and energetic, went into action immediately and, by the end of the day, had obtained equipment to help him lift, move, and reorganize the room. To my amazement, he accomplished the task in less than a week. That's when I learned that the Ehrlich Vineyard barrels from the 2012 and 2013 harvests had survived unscathed. We had dodged a bullet.

In early September, Duffy contacted me to say they were getting ready to bottle the 2012 reds and wanted my approval to put the wine made from our fruit into a vineyard-designated bottle: 2012 B Cellars Ehrlich Vineyard Cabernet Sauvignon. It wasn't a request I was expecting and it both surprised and pleased me. I consulted with my family and we unanimously approved it.

The bottling took place on September 17 while Patricia's brother and sister-in-law, Al and Thu Stankunas, were visiting from Saratoga, California. We all went over to watch. Watching a bottling line in operation is normally the wine industry's equivalent of watching grass grow, but having your name on the bottle changes that. It would be a stretch to say it was exciting, but it was enjoyable. We tasted the wine for the first time while we were there and understood why Duffy had made his request. It was excellent, having the aroma of dark berries and the flavor of berries, tobacco, and spice. Derek Taylor, the associate winemaker, and Jim Borsack gave us a couple of bottles to take home. We stashed them in our wine cellar to be tasted after they had settled and bottle-aged for a few months.

The first third-party validation of its quality occurred when Robert Parker visited the winery in December 2014 and gave it a score of 89-91. He also gave a score of 92-94 to the barrel sample of the 2013. Later the September 2015 issue of *The Wine Enthusiast* gave the 2012 a score of 93 that was accompanied by complimentary words such as richly concentrated, balanced, integrated tannins, and "blackberry and jammy cherry at its core." While credit for the final product has to go to the B Cellars winemaking team, I knew that the decisions the Ehrlich

Vineyard team made on scion and rootstock selection and a variety of farming activities had paid off. As every winemaker knows, great wine starts in the vineyard.

It was while our long-time friends, Joel and Louise Alper, were visiting that we decided to try the "B Cellars Experience." It was September 2015 and the winery had been opened for a year. The tasting area was busy with a dozen other people seated at tables. Jonathan Ruppert and his attentive staff buzzed around taking care of all of us. Our personalized menu described the five-course food and wine pairings that we would be served. The wines ran the gamut from Sauvignon Blanc to Cabernet Sauvignon. The food started with a plate of summer garden relish, basil pesto, California olive oil, and burrata cheese and ended with smoked beef shoulder, a cheddar popover made from scratch, St. Louis BBQ, and charred onion. The quality of the food, the wines, and the service made the experience an exquisite one. The expressions of satisfaction and enjoyment I heard from the tables around us assured me others were having a similar experience. Duffy and Jim were executing the game plan they had explained to me. They were on to something.

The 2015 harvest demonstrated that reversion to the norm is, well … normal. After three plentiful harvests, yields plunged 29 percent from 2014's all-time high. Ehrlich Vineyard experienced a 33 percent drop. Napa Valley Agricultural Commissioner Greg Clark speculated that cool and windy weather in May during bloom led to grape clusters failing to develop properly. The drop was a reminder of the unpredictability of farming.

However, the curtain came down on 2015 in a positive way. Robert Parker re-tasted the 2013 B Cellars Ehrlich Vineyard Cabernet Sauvignon after it had another year of bottle age and gave it a score of 96. He did other things, too, but none more important than this to the owners of Ehrlich Vineyard.

Epilogue

Over the years, Laurie Wood, Nils Venge, Paul Garvey, Wayne Hogue, and a host of other mentors have emphasized the importance of vines being balanced, meaning the number of clusters has to be proportional to the leaves on the canopy. The prevailing wisdom is that if you have too many clusters, you risk growing fruit that won't fully ripen and won't have the rich concentration of flavors winemakers want. If you have too few, you're committing a form of economic suicide.

Patricia and I have spent enough time in the vineyard over the years to know what a balanced vine looks like so, starting in 2012, we have spent hours in the Merlot and Cabernet vineyards attempting to achieve it. With pruning shears in hand, we have gone up and down the rows dropping clusters that were tangled and crowded together, or had developed poorly and were scraggly (a condition called shatter), or would not ripen completely because they were on shoots with too few leaves. It's rewarding because it contributes to the quality of the grapes we grow, which, in turn, contributes to the quality of the wine made from them.

Whether we are out there in the coolness of the morning, listening to birds chirp, or in the heat of early afternoon, hearing the whirr of farm equipment in a neighbor's vineyard, we realize we are less the owners of the land than its stewards and custodians. It was there long before we bought it and it will be there long after we're gone. In the interim, our job is to follow farming practices that preserve it and enable it to be as productive as those practices will allow. It's difficult to imagine a better way to spend time.

Appendix:
Interesting Facts about Wine

What should I do with the cork the waiter puts on the table after opening the bottle of wine I order?

There's no need to do anything with it. The purpose of the bottle-opening ritual is to verify that the wine you ordered is suitable for drinking. The best way to do that is to take the glass the waiter pours wine into, swirl it, put your nose into it, and smell the wine. If there is something wrong with the wine, you'll know it. There usually isn't.

On rare occasion, a bad cork will make the wine undrinkable. There is no mistaking the musty, wet-cardboard smell of a bad cork. Its unpleasantness permeates the air. Your waiter may even catch it before you do.

It is not uncommon for some French wines, particularly red Burgundies, to have a "barnyard" scent. This is different from a bad cork smell. It is caused by a yeast called brettanomyces, or "brett" among oenophiles. It gives the wine what wine writers euphemistically call a "rustic character." It usually dissipates or remains so much in the background that it does not interfere with your enjoyment of the wine. But, at an extreme, it can be overpowering and make the wine undrinkable and you shouldn't accept it. Just explain to the waiter that the wine has been affected by brett. If he understands wine, he'll know what you mean. If he hesitates, just explain it to the maitre'd. Someone will understand.

Why are barrels used in making wine?

Almost all reds and some whites, particularly Chardonnay, are aged in oak barrels that have been "toasted" (charred) on the inside.

Wine in an oak barrel goes through subtle chemical changes. Its porosity allows a very slow oxidation process to occur that enables the wine to soften and evolve. In addition, the chemical composition of the wood – particularly compounds like tannins and vanillins – contributes to the complexity of the wine by creating toasty aromas as well as flavors like vanilla, tobacco, spices, cedar, smoke, and tea.

French oak is preferred because of its high tannin levels and flavors. There are forests in France devoted to growing oak for wine barrels. A barrel made from them currently costs about $1,100. It holds 60 gallons or about 25 cases of wine.

A wine like Cabernet Sauvignon is often aged for 24-30 months in oak, while a Chardonnay can spend 6 to 9 months in oak. A barrel can be re-used for the same grape variety.

Does wine need to "breathe"?

The value of this is widely debated because each wine is different. When the cork is removed, oxygen immediately begins reacting with the wine and any unpleasant aromas have the chance to blow off. However, when you consider the surface area that is exposed to air compared to the volume of wine in the bottle, it's not realistic to think just opening the bottle will make much difference in how the wine will taste.

Rather than just removing the cork, consider letting the wine sit for 10 or 15 minutes to let any unpleasantness dissipate and then pour it into glasses. This action will help aerate the wine and enable you and your guests to enjoy the aromas and textures that develop. There's no need to wait to drink it.

Some wines benefit from decanting, which is pouring the entire contents of the bottle into a decanter. But, if you do this, begin to drink it within 30 to 60 minutes. Wine aromas and flavors can disappear fairly quickly, particularly with old wines.

Is a bottle of Cabernet Sauvignon from California made entirely from Cabernet Sauvignon grapes?

Maybe. For a California wine to be labeled "Cabernet Sauvignon," "Chardonnay," "Pinot Noir," etc. at least 75 percent of the grapes used to make it had to be of that variety. The actual percentage used is much higher than 75 percent in high-end wines.

As the Bordeaux winemakers discovered, the flavor of a red wine can be enhanced if different wine varieties are blended together. High-end California Cabernets often include some Merlot, Cabernet Franc, and Petit Verdot because they add flavor and balance.

Chardonnay tends not to be blended with juice from another grape variety, while Sauvignon Blanc is often blended with Semillion.

By contrast, for "Napa Valley," "Oakville," "Rutherford," or any other California-appellation name to appear on the label, at least 85 percent of the grapes used to make the wine must come from that appellation.

"I get a headache when I drink wine from California but not when I drink wine in Europe. I think it's the sulfites."

Sulfites in wine take the form of sulfur dioxide that occurs naturally during fermentation when it is released by the wine yeasts. Some is added later as a critical component of clean, sound winemaking because it fights bacteria and spoilage.

Congress passed a law during the 1980s that requires all wine sold in the United States, including both domestic and imported, to state on the label that it contains "added sulfites." This requirement is unique to our country. The very same wines sold in their country of origin are not required to contain this statement.

The process of adding sulfur dioxide to wine goes back to the Romans. There is nothing new about it. Using sulfur dioxide in wine production is as common in other parts of the world as it is in the United States.

Sulfites are not unique to wine. They also are found in other fermented products like cheese and beer, as well as in bacon, salami, lunch meats, dried fruits, processed fruit juices, and olives. These products tend to have more sulfur dioxide in them than wine. For example, a 2-ounce serving of dried apricots contains about 110 mg (or 154 ppm) of sulfur dioxide, while a typical glass of wine contains about 10 mg (or 14 ppm) of sulfur dioxide. If you can eat a handful of raisins or dried cranberries without experiencing any problem, it's not likely that sulfites from a glass of wine are the cause of your headaches.

How many calories are there in wine?

There are about 135 calories per 6-ounce glass of wine or about 570 calories per bottle. A useful formula for calculating calories is 1.6 multiplied by the number of ounces consumed multiplied by the alcohol percentage of the wine.

How much wine was created by Jesus at the wedding feast in Cana?

The Gospel According to John, chapter 2 verses 1-11, tells us that six stone jars, each holding 20 to 30 gallons of water, were brought to Jesus. Using the average of 25 gallons per jar, this amounted to 150 gallons, which is equivalent to 570 liters at 3.8 liters per gallon. Since each of today's wine bottles contains .75 liters, the miracle created an amount of wine equal to 760 bottles or slightly more than 63 cases. Let's party!

How should wine scores be interpreted?

Robert Parker's *The Wine Advocate, The Wine Enthusiast,* and *The Wine Spectator* each has its own wine rating system. Each has a procedure for wines to be tasted and evaluated by one or more reviewers on a scale where 100 is the top score. The objective is to convey to their readers how each wine ranks on a comparative scale to wines of a similar variety or type, such as a blend.

Since reference is made in Chapter 29 to how the 2012 B Cellars Ehrlich Vineyard Cabernet Sauvignon was scored by *The Wine Advocate* (89-91) and *The Wine Enthusiast* (93) and how the 2013 B Cellars Ehrlich Vineyard Cabernet Sauvignon was scored by *The Wine Advocate* (96), the following is a description of how those scores can be interpreted:

The Wine Advocate

90-100 is equivalent to an A and is given only for an outstanding or special effort. Wines in this category are the very best produced of their type. There is a big difference between a 90 and 99, but both are top marks. There are few wines that actually make it into this top category because there are not many great wines.

96-100 is an extraordinary wine of profound and complex character displaying all the attributes expected of a classic wine of its variety. Wines of this caliber are worth a special effort to find, purchase, and consume.

90-95 is an outstanding wine of exceptional complexity and character. In short, these are terrific wines.

Wines given scores from 70 to 89 are also reviewed in the publication.

As Mr. Parker points out, "However, there can never be any substitute for your own palate nor any better education than tasting the wine yourself."

The Wine Enthusiast

98-100, Classic: The pinnacle of quality.

94-97, Superb: A great achievement.

90-93, Excellent: Highly recommended.

Wine receiving scores from 80 to 89 will also be rated and may receive a written review.

The Wine Spectator

95-100, Classic: A great wine.

90-94, Outstanding: A wine of superior character and style.

85-89, Very good: A wine with special qualities.

80-84, Good: A solid, well-made wine.

75-79, Mediocre: A drinkable wine that may have minor flaws.

50-74, Not recommended.

Index